30-SECOND
EVOLUTION

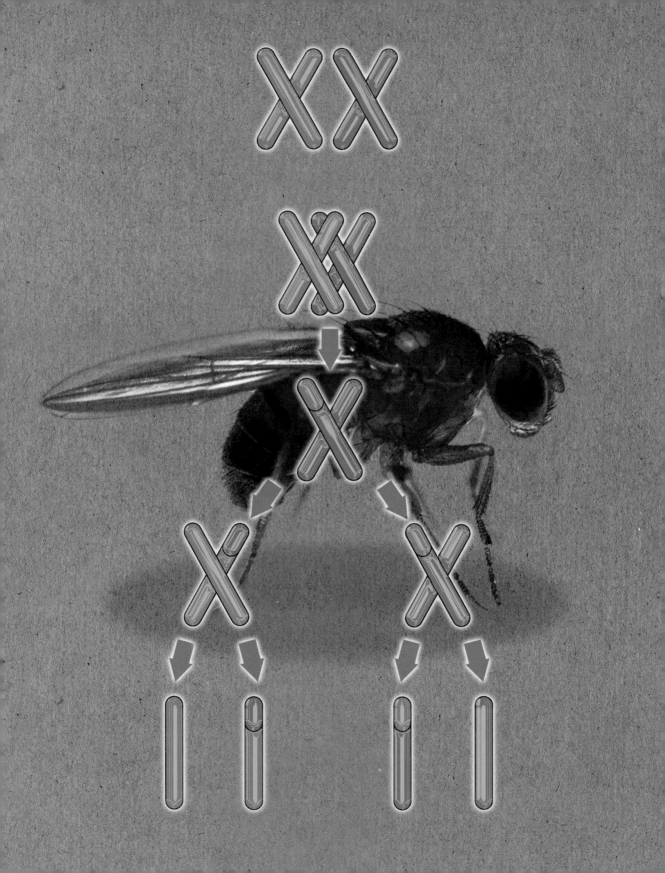

30-SECOND EVOLUTION

The 50 most significant ideas
and events, each explained in
half a minute

Editors
Mark Fellowes
Nicholas Battey

Contributors
Nicholas Battey
Brian Clegg
Isabelle De Groote
Mark Fellowes
Julie Hawkins
Louise Johnson
Ben Neuman
Chris Venditti

ICON

First published in the UK in 2015 by
Icon Books Ltd
Omnibus Business Centre
39–41 North Road, London N7 9DP
email: info@iconbooks.net
www.iconbooks.com

This book was conceived,
designed and produced by

Ivy Press
210 High Street, Lewes,
East Sussex BN7 2NS, U.K.
www.ivypress.co.uk

Creative Director **Peter Bridgewater**
Publisher **Susan Kelly**
Editorial Director **Tom Kitch**
Art Director **Michael Whitehead**
Editor **Jamie Pumfrey**
Designer **Ginny Zeal**
Illustrator **Ivan Hissey**
Glossaries Text **Brian Clegg**

ISBN: 978-1-84831-840-3

Printed and bound in China

Colour origination by
Ivy Press Reprographics

10 9 8 7 6 5 4 3 2 1

CONTENTS

INTRODUCTION
Nicholas Battey & Mark Fellowes

Evolution, caused by the linked processes of natural and sexual selection, accounts for the diversity and interrelationships of all life forms on planet Earth. Although evolution is a theory in the sense that our understanding of it will be modified and developed as scientific understanding grows, it is also much more than that: it is a way of thinking that is fundamental to modern biology and natural history. From the development of language to practical species conservation, evolution is the core concept.

Evolution also accounts for human origins and so conflicts with some religious explanations. It therefore has a colourful history. Charles Darwin's *On the Origin of Species* sparked heated debate and in 1860 Bishop Samuel Wilberforce ('Soapy Sam' for his unguence) asked his debating opponent Thomas Huxley ('Darwin's bulldog') 'Is it on your grandmother's or your grandfather's side that you are descended from an ape?' To which Huxley replied: 'I would rather be descended from an ape than a man who uses his great faculties and influence for the purpose of ridicule.'

More recently the claims of sociobiology – to provide explanations for many aspects of human behaviour based on evolutionary adaptation – have been greeted with fierce opposition from some quarters, while the use of the word 'selfish' by Richard Dawkins in his popularization of evolutionary explanations for altruism added more fuel to the fires of controversy. And eugenics, the idea that human stock can be improved by selective breeding, is an offshoot of evolutionism which has dark associations with the racist programmes of Nazi Germany.

Yet these controversies break out almost exclusively where evolutionary thinking is applied to humans and create a distorted picture of its significance. Evolution addresses the diversity of all life forms – plant, animal, fungal, bacterial and protist. It offers a coherent basis for understanding how the 8.7 million (or perhaps many, many more) species now present on the planet have come into being; through knowledge of population genetics, speciation processes and extinctions it provides not just an explanation of past events but predictions of future chapters in the story of life. These help us understand that every species is a chance event, a combination to be cherished as a never to be repeated expression of the incredible power of evolution. It is evolution acting through natural and sexual selection that has moulded life into the exuberant diversity of form and function that we see around us today.

The evolution of life has depended on one ingredient more than any other: time. The scale of evolutionary time is hard for us to grasp because we naturally think in terms of our own lifespan, the rise and fall of nations and empires – or at most the millennia separating us from the ancient civilizations. Yet evolutionary processes typically occur over millions of years (about 7 million years in the case of the human species) or hundreds of millions of years (dinosaurs of one sort or another were around for 200 million years). The diagram on the following page summarizes the geological eras, periods and epochs referred to extensively in the chapters of this book. The more complex life forms evolved after the Cambrian explosion of life (about 550 million years ago), but about 4,000 million years separate the formation of our planet from the Cambrian epoch, time that was vital to the evolution of the basic elements of life (RNA, DNA, proteins, cells). The diagram also shows the ages of dominance of animal groups, but don't forget that plants were also evolving: during the Carboniferous, vast forests of enormous lycopods, ferns and horsetails flourished; during the Cretaceous, angiosperms (flowering plants) became prominent – as well as the placental mammals.

But this is not to say that evolution acts at this pace, more that this is the scale on which we can see the grand spectacle of evolution unfold. Just as mountains are constantly formed by great tectonic movements and constantly eroded by the elements, evolution and extinction are always at work; we only see the results given the perspective of time. We can, in fact, witness natural selection at work all around us. The evolution of resistance to antibiotics, or of insects to insecticides, is well known. There are many other examples, from the Galápagos finches to the apple maggot fly that show that natural selection is a part of life, and, with time, may give rise to new species. But

GEOLOGICAL TIMESCALE

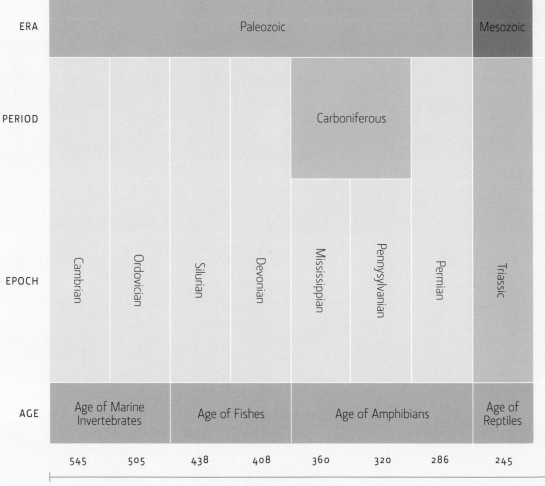

ERA	Paleozoic							Mesozoic
PERIOD					Carboniferous			
EPOCH	Cambrian	Ordovician	Silurian	Devonian	Mississippian	Pennysylvanian	Permian	Triassic
AGE	Age of Marine Invertebrates		Age of Fishes		Age of Amphibians			Age of Reptiles

| 545 | 505 | 438 | 408 | 360 | 320 | 286 | 245 |

Millions of years ago

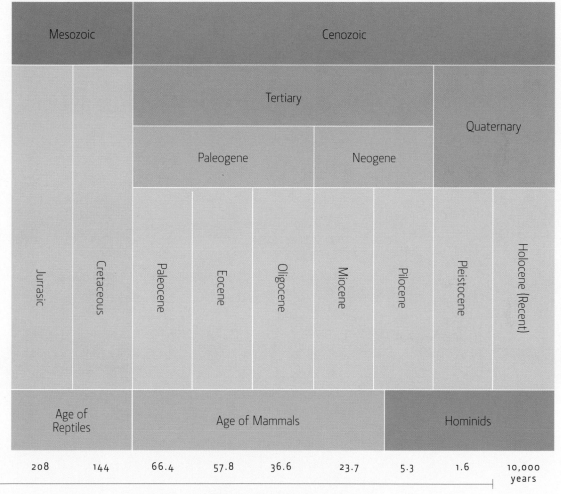

Mesozoic		Cenozoic							

Tertiary

Quaternary

Paleogene

Neogene

Jurasic	Cretaceous	Paleocene	Eocene	Oligocene	Miocene	Pliocene	Pleistocene	Holocene (Recent)

Age of Reptiles	Age of Mammals	Hominids

| 208 | 144 | 66.4 | 57.8 | 36.6 | 23.7 | 5.3 | 1.6 | 10,000 years |

Millions of years ago

hand in hand with evolution goes extinction, and given our own species' malign influence there is no doubt that the creativity of natural selection is now unable to keep pace with our erosion of the Earth's diversity of life.

Our strategy here has been to view evolution from seven different perspectives. In the **History of Evolution** we explore how the modern theory came about, beginning with Darwin's explanation of species origins through natural selection, and incorporating the growing understanding of genes as both agents of stable inheritance and individual variability. In **Origin of Species** modern views of speciation and its genetic underpinnings are described, while **Natural Selection** considers the way genes behave in populations, and populations adapt in response to selection pressure. **Evolutionary History & Extinction** focuses on the geological record and what it tells of life's history; **Evolution in Progress** describes how evolution works, including contemporary examples such as industrial melanism in moths and explanations for apparently non-Darwinian altruism. Sex allows alleles to be exchanged, death is the means of winnowing genotypes; **Sex & Death** looks at how these key events work within an evolutionary framework. The final chapter, **Humans & Evolution,** describes how humans evolved and speculates on our evolutionary future, which paradoxically may involve an escape from natural selection.

Evolutionary thinking has itself evolved – to such an extent that it finds its way into almost all aspects of life. To help capture this diversity each entry in this book is accompanied by a pithy encapsulation (the 3-Second Thrash) and something a bit more speculative – the 3-Minute Thought. This might be a slight mutation of normal thinking, a meme thrown into the mix of ideas that transmit from mind to mind but are rarely, if ever, reproduced as faithfully as genes.

Finally, a thought on how to use this book: dip in, enjoy and be moved to explore further. Life awaits.

THE HISTORY OF EVOLUTION

THE HISTORY OF EVOLUTION
GLOSSARY

Biodiversity The range of animal and plant life within an environment, most commonly based on the number of distinct species.

Divine Creation/Creationism theory
The belief that the Earth and the rest of the Universe, together with all living things in their current forms, are the product of direct supernatural acts of creation by a god or gods, rather than of having evolved through a natural process.

Eugenics Active promotion of the fittest offspring in humans. The English anthropologist and polymath Francis Galton (1822–1911) originally defined it as 'The investigation of... the conditions under which men of a high type are produced.'

Fit/Fittest/Fitness Being well adapted or suited to conditions. In the evolutionary sense 'survival of the fittest' refers to those best suited to survive and pass on genetic material.

Gametes The two kinds of reproductive cell that fuse during fertilization in sexual reproduction: egg cells and sperm cells.

Genetic drift A change in the frequency of a particular variant of a gene that is due not to a selective process but to random fluctuations.

Genotype The genetic instructions that specify what makes an individual organism or cell unique, including variants in parallel sets of chromosomes. Sometimes used to contrast with the phenotype.

Geological epochs/periods/eras
Geologists use units like these to divide up the geological timescale. There are fourteen eras, typically of hundreds of millions of years, divided into periods, which are subdivided into epochs. Periods are probably the most familiar, with names like Cretaceous, Jurassic, Silurian and Cambrian.

Heredity The genetic process by which living organisms pass on characteristics to their offspring (and to the descendants of their offspring). The offspring are said to 'inherit' these characteristics. Genetics is the study of heredity.

Homology A correspondence of an organ or part in its structure or function, whether between the sides of the body or between sexes or species. Reflects a shared genetic ancestry; homology is also used in relation to genes, as well as organs.

Kingdom Originally the highest classification of natural objects (animal, vegetable and mineral kingdoms), but has come to be a taxonomic rank (classification of organisms) that comes above phylum and below domain. The kingdoms are animals, plants, fungi, protists or protoctists and prokaryotes (sometimes divided into archaea and bacteria).

Linnaean classification The taxonomy (classification) of organisms devised by Carl Linnaeus. The modern version of taxonomy has been developed considerably since Linnaeus and now uses kingdoms, classes, orders, families, genera and species.

Phenotype The visible characteristics of a specific organism, sometimes used to contrast with the 'genotype' which is the total set of genetic instructions – the combination of all the genes in a cell or organism.

Punctuated equilibrium An evolutionary theory championed by American biologist Stephen Jay Gould in which species exist for long periods of time with very little evolutionary change, punctuated by relatively rapid events where a species splits into two individual species. The most common alternative, phyletic gradualism, assumes that gradual change eventually results in variants becoming distinct species.

Saltation The idea that a large-scale mutation could produce a new species immediately (the word means a jump or bound). Sometimes confused with punctuated equilibrium, in which species are produced relatively quickly, but still over thousands or tens of thousands of years.

Species A unit of biological classification that is the lowest of the original taxonomic ranks, traditionally defined as a group of organisms capable of interbreeding, although this is not always the case in modern usage. In the traditional two-part name such as *Homo sapiens*, the second word is the species. Animals also may have subspecies, while other kingdoms may have several further sub-divisions.

Stasis Generally meaning inactivity or stagnation, in the punctuated equilibrium theory stasis refers to periods of time with low levels of evolutionary change.

Taxonomy The classification of a set of entities – in biology, of organisms – according to a structured set of principles.

BEFORE EVOLUTION

the 30-second evolution

For John Ray, the 'father of natural history', the world was divinely ordered. Every living thing was designed according to God's plan, from the woodpecker, with its 'short but very strong legs', 'convenient for the climbing of trees'; to leaves that 'concoct and prepare the sap for the nourishment of the fruit, and the whole plant'. In revealing this plan humans would come closer to their maker and to a fuller appreciation of His rationality. There were those who fully supported this approach, including Linnaeus, who founded the modern system of taxonomy in which an organism is classified into species, genus, order, class and kingdom. There were also dissenters, such as Buffon, who believed the Earth to be much older than the 6,000 years derived from the account of creation in the biblical book of *Genesis*. Buffon's idea was that the planets had originated from the Sun and gradually cooled. He put the Earth's age at around 70,000 years, allowing him room to speculate that species had a natural origin. But Buffon was an exception. Even at the beginning of the 19th century, the natural world was generally believed to be populated by creatures fixed at the beginning of time by a divine hand.

3-SECOND THRASH
Everyday experience suggests that species are fixed; according to the Bible they were made by God.

3-MINUTE THOUGHT
Perhaps it's not so silly that people thought species were fixed: after all, in the natural world it isn't obvious that they change; and they are often well designed, both functionally and behaviourally. It took ages for people to work out that the obvious – the Sun goes round the Earth – was wrong. In the same way, the idea that species evolve was counter-intuitive and so met with fierce resistance. Science has a habit of subverting common sense.

RELATED TOPICS
See also
CONTROVERSIES
page 30

SPECIES & TAXONOMY
page 36

3-SECOND BIOGRAPHIES
JOHN RAY
1627–1705
English naturalist and early taxonomist

CAROLUS (CARL) LINNAEUS
1707–78
Swedish founder of modern taxonomy

GEORGES-LOUIS LECLERC, COMTE DE BUFFON
1707–88
French naturalist and author of the multi-volume *Histoire Naturelle*

30-SECOND TEXT
Nick Battey

Most people looked at the Universe and saw the elegantly designed handiwork of a divine creator.

TRANSMUTATION & ARCHETYPES

the 30-second evolution

3-SECOND THRASH
Before Darwin there was the archetype; after Darwin there were ancestors.

3-MINUTE THOUGHT
Mammals as diverse as a bat, a mole and a dolphin have very similar bones making up wing, paw and pectoral fin. This homology was contrasted by Richard Owen with organs that are analogous in terms of function but anatomically different (for example, the wings of bats and birds). This distinction is a lasting legacy from Owen even though his treasured archetype was swept away with the tide of evolutionism.

During the first half of the 19th century a variety of ideas was circulating about the nature and origin of species. In France, leading naturalists were at odds: Lamarck spoke of organisms transmutating – or evolving – into other organisms, while palaeontologist Georges Cuvier denied such changes could occur. In Germany, poet Goethe had a more idealistic vision, of a *bauplan* (blueprint), which underpinned organism development and change. In Britain, Richard Owen in 1848 translated these disparate influences into the concept of the 'archetype' – specifically, the vertebrate archetype. This generalized the vertebrate skeleton into a blueprint used by God to fashion (in succession) the vertebrate species of the world, with humans the latest and closest to perfection. Although the archetype retained a role for a divine Creator, it also allowed for species change, as variations on a theme. In a sense, therefore, it anticipated Darwin's evolutionary theory, published ten years later. But its links to philosophical idealism meant that the archetype had little influence on other scientists. T.H. Huxley, Owen's arch-rival (later Darwin's staunchest supporter), said that it was 'fundamentally opposed to the spirit of modern science'. Yet it set the scene for Darwin to transform ideal archetypes into real ancestors.

RELATED TOPICS
See also
EMERGENCE OF MAJOR
ANIMAL & PLANT GROUPS
page 84

NEW SPECIES
page 106

3-SECOND BIOGRAPHIES
JOHANN WOLFGANG VON GOETHE
1749–1832
German poet who explored the metamorphosis of plants

BARON GEORGES CUVIER
1769–1832
French naturalist at the Musée d'Histoire Naturelle in Paris

RICHARD OWEN
1804–92
British anatomist, founder of the Natural History Museum in London

30-SECOND TEXT
Nick Battey

Richard Owen, developer of the theory of the archetype.

VARIATION & SELECTION

the 30-second evolution

Charles Darwin began his *On the Origin of Species* with a discussion of variation in domestic plants and animals. He had taken up pigeon fancying and had been 'permitted to join two of the London Pigeon Clubs'; in the process he discovered that 'the diversity of breeds is something astonishing'. From the English carrier to the short-faced tumbler; the barb, the pouter, the turbit, the Jacobin, the trumpeter and laugher: all these pigeon breeds differed in truly remarkable ways. Yet, Darwin argued, they were all derived from a single wild species, the rock pigeon (*Columba livia*). How had this happened? During the process of domestication humans had picked out particular individuals: 'The key is man's power of accumulative selection: nature gives successive variations; man adds them up in certain directions useful to him. In this sense he may be said to make for himself useful breeds.' By analogy, in the wild individuals of a species vary, but here nature is the selector instead of a human. Competition for resources means that the best-adapted individuals flourish – in other words, they are selected. In this way new species are created in response to changing environments and circumstances. *Natural* selection was Darwin's crucial idea, his decisive insight into the mechanism behind evolution.

3-SECOND THRASH
At the heart of Darwin's theory of evolution is the idea that variation can be enhanced by selection.

3-MINUTE THOUGHT
Darwin anticipated the resistance his theory would encounter, particularly amongst experienced naturalists used to thinking in terms of a 'plan of creation'. He placed his hope in the younger generation who could be more impartial and might accept a theory that succeeded in explaining some of the facts, despite leaving others unanswered. Now it seems that optimism was justified; yet it took about 80 years for the idea of natural selection to be fully accepted.

RELATED TOPICS
See also
THE MODERN SYNTHESIS
page 28

FROM ADAPTATION TO
SPECIATION
page 48

TYPES OF SELECTION
page 68

3-SECOND BIOGRAPHIES
CHARLES DARWIN
1809–92
British naturalist and co-inventor of the theory of evolution by natural selection (1858). *On the Origin of Species By Means of Natural Selection* (1859) set out the theory in detail

ALFRED RUSSEL WALLACE
1823–1913
British naturalist and co-inventor of the theory of evolution by natural selection

30-SECOND TEXT
Nick Battey

Darwin realized that a version of selective (artificial) breeding takes place in the wild – natural selection.

12 February 1809
Born in Shrewsbury to Robert and Susanna Darwin

1831
Receives ordinary degree from Cambridge

27 December 1831
Departure of the HMS *Beagle* from Plymouth

15 September 1835
Beagle expedition reaches the Galapagos Islands, off South America

2 October 1836
Beagle returns to the UK, landing at Falmouth

24 January 1839
Darwin elected to Fellowship of the Royal Society

29 January 1839
Darwin marries his cousin, Emma Wedgwood – they will have 10 children between 1839 and 1856

18 June 1858
Darwin receives paper on natural selection from Alfred Russel Wallace

1 July 1858
Darwin and Wallace's material is jointly presented to the Linnean Society

24 November 1859
Publication of *On The Origin of Species*

30 June 1860
Evolution debate at the Museum of Natural History, Oxford

24 February 1871
Publication of *The Descent of Man*

19 April 1882
Dies of heart disease at Down House, Kent

2002
BBC conducts a poll to determine the '100 greatest Britons'. Charles Darwin was number 4, ahead of any other scientist

CHARLES DARWIN

Charles Darwin's family intended Darwin to be a doctor, and sent him off to medical school at Edinburgh, but after two mediocre years he transferred to Cambridge to take a BA, aimed at a career in the Church of England. Here a growing interest in beetle-collecting and a friendship with a botany professor stimulated him more than his theological studies.

During his last months at Cambridge, Darwin sat in on a geology course, which helped when his professor friend, John Henslow, recommended Darwin for the journey that transformed his life. He was asked to be companion to Robert Fitzroy, captain of the HMS *Beagle* on a voyage mapping the coastline of South America. During this five-year trip, returning via Australia and the Cape, it was probably geology that kept Darwin busiest, but he also made extensive collections of plant and animal specimens and studied everything from plankton to fossilized mammoth bones.

Most significant for his later theories was Darwin's visit to the Galapagos archipelago, where the variation in birds and tortoises from island to island led him to speculate that animal species were not as fixed in creation as was assumed. These thoughts were sidelined on his return, but by the following year Darwin was making notes suggesting that one species could change into another, and that the descent of species took the form of a branching tree.

Darwin did not rush to publish his ideas. It was only when he received a letter from British naturalist Alfred Russel Wallace, 22 years after the *Beagle*'s return, that Darwin changed pace. By then, he had started writing a book, but Wallace's letter outlined an almost identical theory of natural selection. Despite technically having precedence, Wallace did not press the point and their work had a joint presentation to the Linnean Society in July 1858. This drew little attention, unlike Darwin's book, *On the Origin of Species by Natural Selection or the Preservation of Favoured Races in the Struggle for Life*, which sold out on publication. Darwin went on to bring humans into the picture in his 1871 *The Descent of Man, and Selection in Relation to Sex*.

Darwin wrote a number of other books, but his place in history was won by that theory which now seems so inevitable. He died at his Kent home in 1882.

THE REDISCOVERY OF MENDEL

the 30-second evolution

Variation, the raw material on which natural selection acts, was key to Darwin's theory. But where did it come from and how was it transmitted down generations? Finding an answer took decades, but a significant step was taken by Gregor Mendel. He discovered that discrete characteristics of pea plants (for example, tallness or shortness) were passed to the next generation in a consistent way. Each characteristic was determined by one factor inherited from each parent, and one factor always dominated over the other (for example, smooth over wrinkly pea skin). Mendel's results, published in 1866, were ignored until 1900, by which time the founders of the new discipline of genetics had concluded that big changes in discrete characters ('saltations') were responsible for the origin of new species, rather than natural selection acting gradually on small variations. These 'saltationists' seized on Mendel's findings as evidence supporting their mechanism of evolution. This meant that although the idea of evolution by descent from a common ancestor was accepted quite quickly after publication of *On the Origin of Species* in 1859, Darwin's mechanism of natural selection was challenged. Only in the 1930s were Mendel's laws conclusively shown to be compatible with Darwin's way of thinking.

RELATED TOPICS
See also
UNDERSTANDING GENES IN POPULATIONS
page 26

MUTATION & SPECIATION
page 46

GENETIC VARIATION
page 64

3-SECOND BIOGRAPHIES
GREGOR MENDEL
1822–84
Austrian monk and discoverer of the laws of inheritance

WILLIAM BATESON
1861–1926
British biologist and champion of Mendel who coined the term genetics in 1905

30-SECOND TEXT
Nick Battey

3-SECOND THRASH
Darwin saw the importance of variation; Mendel helped us understand it by showing how characters are transmitted across generations.

3-MINUTE THOUGHT
Mendel, an amateur, chose peas that would breed true for the characters he was interested in, and worked systematically to obtain clear-cut results. He sent his findings to Carl von Nageli, a professional scientist, who was critical and unhelpful, suggesting Mendel work on hawkweed; but this species has an unusual mode of reproduction, in which the male rarely contributes to the next generation. Mendel therefore could not generalize his results and died with his work unrecognized.

Mendel had trained in maths and his mathematical knowledge enabled him to express his experimental findings clearly.

UNDERSTANDING GENES IN POPULATIONS

the 30-second evolution

3-SECOND THRASH
Quantitative genetics
provided a solid basis
for evolution by
natural selection.

3-MINUTE THOUGHT
T.H. Morgan was
forthright: in 1905 he
rejected Darwin's
mechanism for the origin
of species: 'new species
are born; they are not
made by Darwinian
methods, and the theory
of natural selection has
nothing to do with the
origin of species, but with
the survival of already
formed species.' At this
time Morgan envisaged
species arising through
saltations – major,
large-scale changes. But as
the evidence accumulated,
he changed his views and
accepted Darwinian natural
selection, deferring to
experimental findings.

After the rediscovery of Mendel's work in 1900, major developments occurred in genetics, spearheaded by T.H. Morgan's group in the United States. Working with the fruit fly *Drosophila melanogaster* they confirmed that regions of the chromosomes ('genes') were the units of heredity. The way chromosomes behaved during the formation of the male and female gametes (sperm and egg) meant each contained a mixture of paternal and maternal characters. This gave a physical basis for Mendel's description of the discrete transmission of characters across generations according to clearly defined rules. The group also analysed linkage and recombination effects, and identified mutation as the mechanism by which new variation was introduced into genes. Morgan's research gave a mechanistic foundation for heredity, a contribution to science for which he received the Nobel Prize in 1933. Meanwhile, R.A. Fisher, J.B.S. Haldane and Sewall Wright used mathematics to describe the behaviour of genes in populations, and showed that Mendel's laws could be reconciled with Darwinian natural selection. Many traits that varied continuously rather than discretely could be explained on the basis of large numbers of genes acting together and segregating in Mendelian fashion. Thus quantitative approaches to genes in populations helped link genetics with evolution.

RELATED TOPICS
See also
THE REDISCOVERY OF
MENDEL
page 24

GENES
page 60

GENETIC VARIATION
page 64

3-SECOND BIOGRAPHIES
T.H. MORGAN
1866–1945
American pioneer of genetics
and key figure in the
establishment of the physical
basis of heredity

R.A. FISHER
1890–1962
British mathematician; his
1930 book *The Genetical
Theory of Natural Selection*
explored how natural selection
acted on genes in populations

30-SECOND TEXT
Nick Battey

T.H. Morgan's experiments with fruit flies demonstrated how genes behaved, and the key role of mutation.

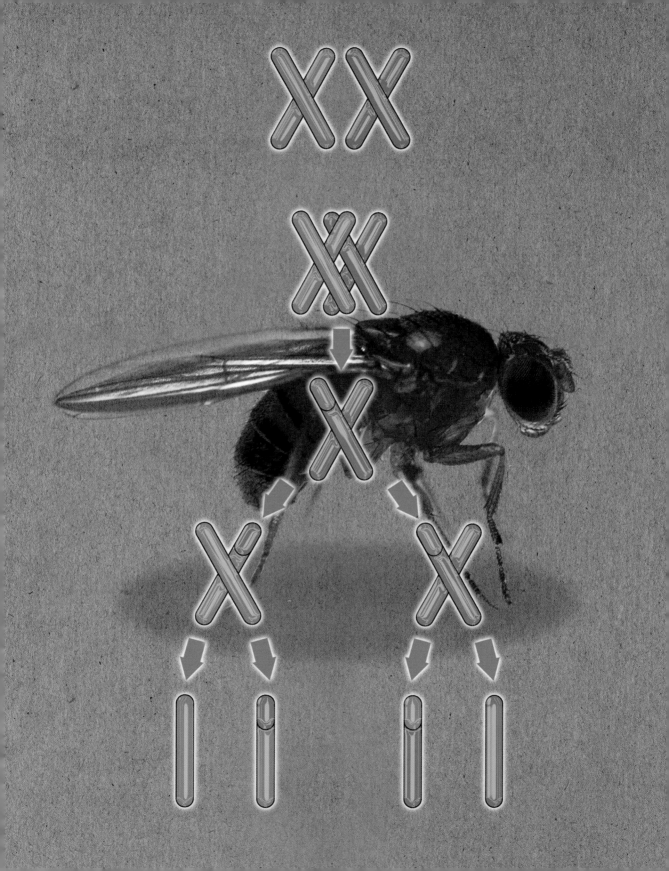

THE MODERN SYNTHESIS

the 30-second evolution

The 'modern synthesis' combined genetics and natural selection. The phrase is from Julian Huxley's book *Evolution: The Modern Synthesis* (1942), in which he emphasized the compatibility of Mendelism and Darwinian evolution. The challenges were to explain the fit (or adaptedness) of organisms to their environments, which results in the amazing spectrum of biodiversity and the fact that while every individual is different, individuals group into naturally distinct units – species. How species formed and the relative importance to speciation of geographic isolation, ecological specialization and genetic divergence were analysed by Huxley. The book reflected the spirit of the time, in which broad agreement on the mechanisms of the evolutionary process had been reached. Uncertainties remained: was the unit of selection the gene, the individual or the population? Was evolution gradual or did the fossil record indicate rapid speciation followed by long periods of stasis ('punctuated equilibrium')? Did adaptation to their environment always shape species or was much evolution neutral, due to genetic drift? How should evolutionary thinking be applied to humans, with the complicating influence of culture? But consensus had been reached on a scientific approach that came to dominate biology and wider thinking about life on Earth.

3-SECOND THRASH
Darwin's ideas gained wide support through the unification of evolutionary biology and genetics.

3-MINUTE THOUGHT
Is evolution progressive? Huxley concludes: 'the demonstration of the existence of a general trend which can be legitimately be called progress, and the definition of its limitations, will remain as a fundamental contribution of evolutionary biology to human thought.' J.B.S. Haldane, in *The Causes of Evolution* (1932), to which Huxley acknowledges considerable debt, differs: 'when we speak of progress in Evolution we are already leaving the relatively firm ground of scientific objectivity for the shifting morass of human values.' Around human values cluster the controversies associated with evolution.

RELATED TOPICS
See also
UNDERSTANDING GENES IN POPULATIONS
page 26

CONTROVERSIES
page 30

EVOLUTIONARY RATES & EXTINCTIONS
page 86

3-SECOND BIOGRAPHIES
THEODOSIUS DOBZHANSKY
1900–75
Ukranian-American geneticist who helped build the modern evolutionary synthesis

JULIAN HUXLEY
1887–1975
British zoologist, wildlife conservationist and eugenicist

SEWALL WRIGHT
1889–1988
American pioneer of evolutionary genetics

30-SECOND TEXT
Nick Battey

Julian Huxley was the grandson of Darwin's great supporter, T.H. Huxley.

CONTROVERSIES
the 30-second evolution

Evolution is a creation story based on empirical evidence. It draws together observations from fossils, present-day species distributions, similarities in DNA sequence and morphology. It enables us to understand the diversity of life and offers an explanation for who we are and where we came from. Because of its wide scope and deep significance, evolution challenges other views. Until Darwin, people could accept stories about origins that were based on tradition: the creationism theory. Since Darwin, such myths have become difficult to defend. Yet there is still variation in acceptance of evolution: in the UK by about two-thirds of the population; in the US, by under half. Another controversial issue related to evolution is eugenics. Eugenics advocates the improvement of the human race through regulated breeding, an idea many biologists in the late 19th and early 20th centuries supported; for instance, by arguing for increased reproduction in the upper classes, and its discouragement in the poor. Some US states had programmes of sterilization of 'degenerates'. But extreme persecution of Aryan 'defectives' and Jews in Nazi Germany discredited the eugenics movement. Post-Second World War, research was redirected into human and medical genetics, an area that remains controversial because of issues of choice and personal belief.

RELATED TOPICS
See also
BEFORE EVOLUTION
page 16

HUMAN EVOLUTION:
THE FUTURE
page 152

3-SECOND BIOGRAPHIES
FRANCIS GALTON
1822–1911
British polymath and promoter of eugenics

CHARLES DAVENPORT
1866–1944
American eugenicist and director of the Cold Spring Harbor Laboratory in New York

30-SECOND TEXT
Nick Battey

Early eugenicists used pedigree charts to improve genetic qualities and in extreme cases, such as Nazi Germany to track Jewish ancestry. Modern pedigree charts show inheritance of specific traits (such as baldness).

3-SECOND THRASH
Evolution is controversial because it challenges our personal beliefs and assumptions.

3-MINUTE THOUGHT
Genotype influences human development, disease, behaviour, intelligence, personality – in fact, all aspects of our lives. We now have the power to change genotype through medical genetics. An evolutionary perspective shows our future is not predetermined; but the history of eugenics says 'proceed with care'. Charles Davenport and others fuelled faulty science with contemporary prejudice in an attempt to eliminate human variation, that most valuable of evolutionary commodities.

THE ORIGIN OF SPECIES

THE ORIGIN OF SPECIES
GLOSSARY

Asexual reproduction Reproduction in which an organism only has a single parent and thus is a genetic copy or clone of that parent, carrying only genes from that parent. Includes parthenogenesis, fission (splitting into two), sporulation and fragmentation.

Class Originally the subdivisions of the animal, vegetable and mineral kingdoms, class is now a grouping of organisms, a taxonomic rank that comes above order and below phylum or division. Mammals, for example, form a class.

Genotype The genetic instructions that specify what makes an individual organism or cell unique, including variants in parallel sets of chromosomes. Sometimes used to contrast with the phenotype.

Genus A group of organisms that possesses common characteristics that make it distinct from those of any other group, the genus is a taxonomic rank (classification of organisms) that comes above species and below family. In the familiar two-word Latin naming structure (for example, *Homo sapiens*), the first word is the genus, the second the species.

Hybridization The formation of hybrids, organisms that are cross-bred between parents, usually of different species. Hybridization can also take place between subspecies, genera or occasionally families.

Kingdom Originally the highest classification of natural objects (animal, vegetable and mineral kingdoms), but has come to be a taxonomic rank (classification of organisms) that comes above phylum and below domain. The kingdoms are animals, plants, fungi, protists or protoctists and prokaryotes (sometimes divided into archaea and bacteria).

Lateral gene transfer Transferring genes between organisms other than via reproduction.

Linnaean classification The taxonomy (classification) of organisms devised by Carl Linnaeus, which incorporated kingdoms, classes, orders, families, genera and species. The modern version of taxonomy has been developed considerably since Linnaeus.

Mutation A change in the genetic material of an individual that may be passed on to the next generation. Mutations can, but do not necessarily, cause changes in the phenotype.

Order A distinction in the taxonomy (classification) of organisms that falls between class and family. Primates and Lepidoptera are example of orders.

Phenotype The visible characteristics of a specific organism, sometimes used to contrast with the 'genotype' which is the total set of genetic instructions – the combination of all the genes in a cell or organism.

Phylogeny/phylogenetic tree A branching diagram – sometimes called a 'tree of life' – that shows evolutionary links between species. These were originally based on physical characteristics, but are now more likely to be dependent on genetic similarity. Phyla (sing. phylum) are a taxonomic division below kingdom and above class.

Species A unit of biological classification that is the lowest of the original taxonomic ranks, traditionally defined as a group of organisms capable of interbreeding, although this is not always the case in modern usage. In the traditional two-part name such as *Homo sapiens*, the second word is the species. Animals also may have subspecies, while other kingdoms may have several further sub-divisions.

Taxa The plural of 'taxon', the set of organisms in a particular unit such as a species. The basic component in a structured (taxonomic) classification such as the Linnaean system.

Taxonomic impediment The limitations of our knowledge of the taxonomy of living things on Earth. The fact that the majority of species are probably not yet properly classified (and that there is a shortage of taxonomists to undertake this task) is sometimes described as the taxonomic impediment.

Taxonomists Someone involved in classification – in biology, typically an expert in the classification of living organisms.

Taxonomy The classification of a set of entities – in biology, of organisms – according to a structured set of principles.

SPECIES & TAXONOMY

the 30-second evolution

3-SECOND THRASH
Taxonomists are striving to document species and build classifications that reflect the evolutionary trajectory of species before these species become extinct.

3-MINUTE THOUGHT
The fundamental unit of Linnaean classification is the species. It is disputed whether species are 'real', and the entities that evolution acts upon, or arbitrary creations of taxonomists, like the higher-ranked groups. Real or not, recognizing and describing species is fraught with problems. One aspect of the species problem is that there is no universally applicable species concept that all taxonomists can adopt.

Chapter XIV in Darwin's *On the Origin of Species* begins 'From the most remote period in the history of the world organic beings have been found to resemble each other in descending degrees, so that they can be classed in groups under groups.' This nested hierarchy of groups within groups was apparent in the 18th-century classification system of Linnaeus – the system of classification used in Darwin's time and still today. The job of building these classifications, and modifying them to reflect new findings, falls on taxonomists. Their Herculean task is to discover and describe all living species, to arrange them into nested groups and to give each group a rank, all while following complex rules on naming. There may be in the region of 9 million eukaryotic species (organisms whose cells contain a nucleus), and less than a quarter catalogued. We rely on taxonomic information to conserve, manage and use biodiversity. The worldwide shortage of this taxonomic information is known as the taxonomic impediment. Taxonomists protest that although humans have walked on the Moon, sought life on Mars and sequenced the human genome, the urgent task of documenting the species with which we share our planet is neglected.

RELATED TOPICS
See also
CHARLES DARWIN
page 22

BUILDING PHYLOGENIES
page 38

3-SECOND BIOGRAPHIES
CARL LINNAEUS
1707–78
Swedish botanist who outlined a nested hierarchy and established universally accepted conventions for naming organisms

CHARLES DARWIN
1809–82
English naturalist whose theory explained why natural classifications are comprised of groups within groups.

30-SECOND TEXT
Julie Hawkins

Carl Linnaeus, father of taxonomy, rose from humble beginnings to be one of Europe's greatest scientists. Philosopher Jean-Jacques Rousseau said, 'I know of no greater man on Earth'.

BUILDING PHYLOGENIES

the 30-second evolution

In 1837 Darwin drew a tree-like diagram in his notebook, above which he wrote: '*I think*'. He believed his diagram represented the expected result of descent with modification – the evolutionary process itself. Darwin's diagram was among the first depictions of what we now call a phylogeny. Phylogenies, or phylogenetic trees, portray how, over millions of years, ancestral species give way to new descendant species. They therefore describe how species are related to each other. Thus, phylogenies also tell us about the interrelatedness of more inclusive taxonomic groups such as genera, families or orders. After all, the literal meaning of the word *phylo-geny* is genesis, or origin, of groups. Historically, phylogenies were built from analyses of the similarities and differences between species on a suite of morphological characteristics. Today, genetic data have supplanted morphological characters as the data of choice to build phylogenies. Gene-sequence data can provide a more accurate view of how species or groups are related – the use of these data has changed the way we think about how many species have evolved. For example, until the 1990s scientists believed that hippopotamuses were closely related to pigs. However, using phylogenies built with gene-sequence data we now know they are actually far closer to whales and dolphins.

3-SECOND THRASH
Phylogenies provide a realistic and intuitive view of how descendant species are produced from ancestral species.

3-MINUTE THOUGHT
Humans are more closely related to chimpanzees than to gorillas. The reason for this is because some 5–10 million years ago humans and chimpanzees shared a common ancestor that gave rise to both species. However, interspecific hybridization and lateral gene transfer are rife in some organisms (especially plants and bacteria). It is interesting to think about what it would mean to build a phylogenetic tree for these sorts of species.

RELATED TOPICS
See also
SPECIES & TAXONOMY
page 36

MUTATION & SPECIATION
page 46

3-SECOND BIOGRAPHY
CHARLES DARWIN
1809–82
English naturalist and geologist

30-SECOND TEXT
Chris Venditti

Darwin's sketch pioneered phylogeny ... modern scientists used gene-sequence data to prove that hippos are not closely related to pigs.

MAKING SPECIES: ISOLATION

the 30-second evolution

Speciation, or the origin of new species, requires reproductive isolation. It is easy to imagine how the environment can impose isolation between two incipient species – a river or a mountain range might geographically separate two populations such that they cannot reproduce. However, for the process of speciation to be completed, genetically based barriers to gene-flow must evolve, such that if the populations came into contact again no viable offspring could be produced. There are several ways this process can occur. Pre-mating isolating mechanisms prevent mating. In some snail species, for example, the direction in which the shell coils is determined by a single gene – this, coupled with the fact that species with right-coiling shells cannot physically mate with those that are left-coiling, means interbreeding is impossible. Similarly, genetically based temporal isolation can inhibit mating. This is the case in two closely related *Drosophila* (fruit fly) species in which one breeds early in the day while the other breeds in the afternoon. In some situations mating can occur but genetically evolved differences mean that the egg is not fertilized or the resultant offspring is sterile. An example of the latter is the mule or hinny, the sterile progeny of a horse and a donkey. These mechanisms are known as 'post-mating isolating mechanisms'.

RELATED TOPICS
See also
SPECIES & TAXONOMY
page 36

MECHANISMS OF ISOLATION
page 44

MUTATION & SPECIATION
page 46

FROM ADAPTATION TO SPECIATION
page 48

30-SECOND TEXT
Chris Venditti

3-SECOND THRASH
Species are made when two populations are separated and genetically based differences occur that mean a viable offspring cannot be produced.

3-MINUTE THOUGHT
We all develop an intuitive impression of what a species is – a lion is not a tiger – and biologists have a good understanding of how new species are formed. However, things can get more complicated in organisms that do not reproduce sexually. In asexually reproducing species offspring are genetically identical to their parents and each other. How could speciation occur in these organisms?

Snails with opposite-coiling shells cannot mate, while the hinny, offspring of a horse and a donkey, is usually sterile – it cannot reproduce.

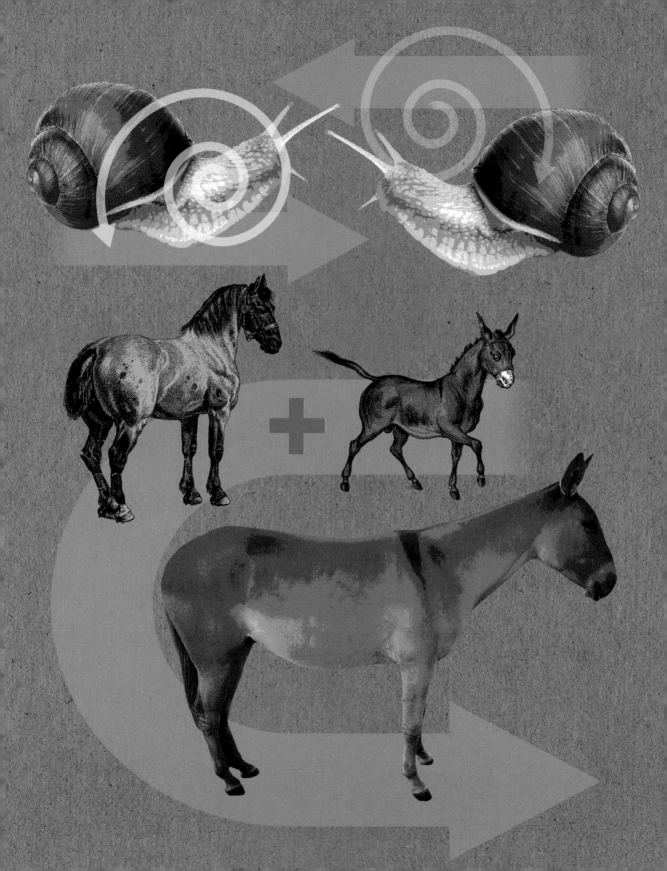

8 January 1823
Born in Llanbadoc to Thomas and Mary Wallace

1828
Family move to Hertford, where Wallace attends grammar school

1837
Moves to London, then follows his brother to Kington and Neath, training to be a surveyor

1848
Leaves for Brazil on the *Mischief* with entomologist Henry Bates

1852
Travels back to the UK on the *Helen*, which is abandoned 26 days into the voyage

1854
Begins eight-year expedition to the East Indies, collecting more than 125,000 specimens

18 June 1858
Wallace's paper on natural selection received by Charles Darwin

1 July 1858
Darwin and Wallace's work presented to the Linnean Society

1866
Marries Annie Mitten – they will have three children

1869
Publishes his book *The Malay Archipelago*

1881
British government awards Wallace £200 annual pension

7 November 1913
Dies at his home, Old Orchard, at Broadstone in Dorset

ALFRED RUSSEL WALLACE

While almost everyone knows of Charles Darwin, far fewer are aware of the parallel development of evolutionary theory by Alfred Russel Wallace. Born to an English family in Llanbadoc, a village in southeast Wales, Wallace moved to his mother's hometown, Hertford, when he was five. Owing to financial difficulties, his education was interrupted when he was thirteen, after which he picked up his knowledge on the job, initially working as a surveyor, apprenticed to his oldest brother, William. When work dried up, Wallace was able to use his experience to secure a position as a teacher at the Leicester Collegiate School.

Like Darwin, Wallace developed an interest in collecting insects, an occupation that fitted well with the surveying work he returned to after leaving the school. Following a brief spell lecturing at the Neath Mechanics' Institute, Wallace joined up with his Leicester entomologist friend, Henry Bates, to form an expedition to Brazil, collecting specimens for sale and enabling Wallace to look for evidence to support his ideas on the transmutation of species. The voyage lasted four years, but on the return trip the ship caught fire, destroying Wallace's specimens and leaving him stranded with the crew in an open boat for ten days.

Two years later, Wallace embarked on an eight-year expedition to the East Indies lasting until 1862, collecting many specimens and discovering thousands of previously unknown species. During his travels he refined his theory of evolution by natural selection, writing a number of papers culminating in a letter sent to Darwin in 1858, outlining his ideas and triggering Darwin's completion of *On the Origin of Species*. Thanks to Darwin, Wallace's work was made public, though it was soon eclipsed by Darwin's book. It is impossible not to admire Wallace's generosity of spirit, writing in 1860: 'Mr Darwin had given the world a *new science*, and his name should, in my opinion, stand above that of every philosopher of ancient or modern times. The force of admiration can no further go!!!'

It wasn't until 1866 that Wallace married. For a period his finances were unstable with heavy losses on bad investments and despite his expertise and experience, Wallace was unable to get a permanent position. Thanks to Darwin, though, he was awarded a government pension, which with his income from writing helped keep him solvent. He died at the age of 90 and was buried in Dorset.

Brian Clegg

MECHANISMS OF ISOLATION

the 30-second evolution

Biologists and science-fiction
writers have been interested in hybrids, real and
imagined, for a long time. How about the zorse
(zebra x horse) or liger (lion x tiger)? Or
humanzees, chumans or manpanzees? Some
species pairs are freely fertile, such as the lion
and tiger, but hybrids exist only in captivity
because the distributions of the parental species
do not overlap. Other pairs don't hybridize
because of pre-fertilization barriers to
interbreeding. For example, many bird species
have evolved strikingly different sexual displays
and sexual ornaments and won't select mates
without these identifiers. Or closely related
flowering plants use different floral displays to
attract different pollinators, or flower at different
times of day, resulting in reproductive isolation. If
these pre-fertilization barriers are overcome, it
may be because humans have intervened. In *On
the Origin of Species*, Darwin wrote about the
work of Kölreuter and Gärtner, who carried out
hand-pollinations, the manual transfer of pollen
from one plant species to another. Sometimes
after experimental fertilization, especially if the
parents are distant relatives, fertile offspring fail
to develop. Ongoing scientific study has
explained post-fertilization mechanisms of
reproductive isolation, revealing the genetic
basis of incompatibility.

3-SECOND BIOGRAPHIES
JOSEPH GOTTLIEB KÖLREUTER
1733–1806
German botanist who
pioneered the study of
hybridization to understand
the origin of species, creating
the first scientific hybrid
between tobacco species

CARL FRIEDRICH VON
GÄRTNER
1772–1850
German botanist and the
wealthy son of a professor of
botany at St Petersburg;
devoted staff, grounds and
finances to lifelong study of
plant hybridization

GEORGE LEDYARD STEBBINS
1906–2000
American botanist who
published *Variation and
Evolution in Plants* (1950), an
influential text contributing to
the modern evolutionary
synthesis, and including an
important chapter on isolation

30-SECOND TEXT
Julie Hawkins

*Barriers can be
overcome to produce
some extraordinary
hybrids, but we
won't find the
fantastical hippogriff
(horse x eagle).*

MUTATION & SPECIATION

the 30-second evolution

DNA contains all the information needed to build an organism. It dictates how it will look, function and behave. A mutation is a natural and relatively common event that changes an organism's DNA during duplication in cell division – it's basically a mistake! However, mutations are exceptionally important mistakes as they form the raw material on which evolution by natural selection works – they can cause changes to the phenotype. Mutations in reproductive cells (eggs and sperm) are passed to the next generation. The effect of a mutation is variable, some can be beneficial, some can be negative while others produce no effect at all. Mutation is the origin of variation in a population. It is this variation that natural selection acts on to sculpt a species to be better able to survive and reproduce in a given environment. If a mutation causes a beneficial change the individual who carries it will be better equipped to survive and reproduce and therefore the mutation will become predominant in that population. If two populations of a species are isolated it is easy to see how a build-up of mutations in isolation can eventually lead to speciation.

3-SECOND THRASH
Mutations explain why individuals are distinct from one another and form the basis for evolution by natural selection.

3-MINUTE THOUGHT
Sometimes a mutation can occur in a particular specialized gene that has a significant effect on the phenotype of an organism. One famous example is the formation of legs on the head of a fruit fly where antennae are expected. Such mutations are often catastrophic, causing the quick demise of the individual carrying it. But, some argue that such mutations can be beneficial and instantaneously result in a new species or evolutionary lineage; such a mutation has been posited for the origin of turtles.

RELATED TOPICS
See also
VARIATION & SELECTION
page 20

MAKING SPECIES: ISOLATION
page 40

MECHANISMS OF ISOLATION
page 44

FROM ADAPTATION TO SPECIATION
page 48

3-SECOND BIOGRAPHY
ETIENNE GEOFFROY
SAINT-HILAIRE
1772–1844
French zoologist whose work on fossil amphibians and reptiles led him to posit that transmutations might occur rapidly, not gradually

30-SECOND TEXT
Chris Venditti

The mutation that caused legs to grow on a fruit fly's head had no benefits, but some mutations result in evolutionary progress.

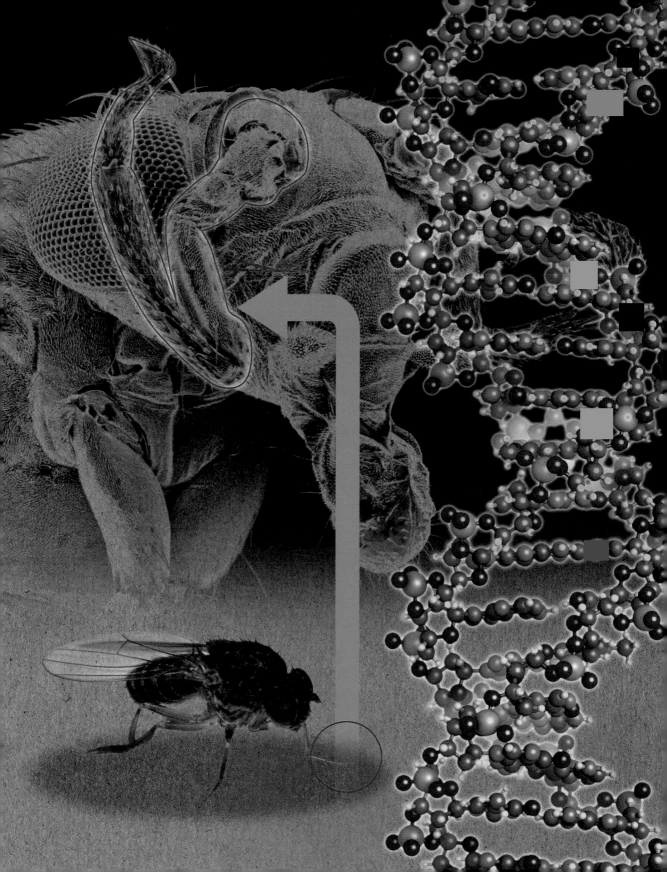

FROM ADAPTATION TO SPECIATION

the 30-second evolution

Adaptation is implicit in 'survival of the fittest'. The Aye-aye is a nocturnal primate endemic to Madagascar. It has a very long thin middle finger that it uses to tap on trees to locate grubs beneath the bark surface. Once located, it uses that finger to tease out the grub. This finger is an example of an adaptation. Over the course of time the process of natural selection used variation caused by mutation to sculpt the Aye-aye's finger. This adaptation is now shared by all Aye-ayes and contributes to the fitness and survival of individuals. In general an adaptation is any characteristic that facilitates the survival or reproduction of an organism that has it. Bird migration, cactus spines and leopard spots are three of the billions of adaptations that organisms exhibit today; there have been billions more in species now extinct. Adaptation can lead to divergence between populations and eventually speciation. If two isolated populations of a species are exposed to different selective pressures we would expect natural selection to result in varying adaptations between the two. Over time, these populations may become so different that if they were to come in contact again they would not be able to mate or produce viable offspring; thus speciation would have occurred.

RELATED TOPICS
See also
VARIATION & SELECTION
page 20

MAKING SPECIES: ISOLATION
page 40

MECHANISMS OF ISOLATION
page 44

MUTATION & SPECIATION
page 46

3-SECOND BIOGRAPHIES
CHARLES DARWIN
1809–82
English naturalist and geologist

30-SECOND TEXT
Chris Venditti

3-SECOND THRASH
Anything that facilitates an organism's ability to survive or reproduce is an adaptation.

3-MINUTE THOUGHT
When one starts to consider species' characteristics it is hard to think of something that is not an adaptation. However, we and other organisms possess many characteristics that are not adaptations. Some characteristics are by-products of others. For example, our blood is red, but not by virtue of the fact that a human with orange blood would be less able to survive or reproduce. It is red owing to the chemical composition of our blood.

Aye-ayes are solitary: they hunt alone by night in the forests of Madagascar, using their long middle finger – a result of adaptation – to tease out bugs.

SPECIES DIVERSITY

the 30-second evolution

3-SECOND THRASH
The millions of species that exist today are not randomly distributed across Earth: far more species occur nearer the equator than at the poles.

3-MINUTE THOUGHT
The idea that the distribution of species is driven by the balance between speciation and extinction is an interesting one. It accords with studies that have shown that some habitats promote reproductive isolation more than others and therefore have the potential to change the tempo of speciation. If habitats that promote reproductive isolation mirror (or oppose) the latitude–species diversity pattern, this might help pinpoint a mechanism.

An estimated 8.7 million species inhabit every corner of our planet – even the harshest environments. There are frogs that freeze solid then spring back to life once thawed, worms that can thrive in water that is almost at boiling point and even tiny organisms that live in rocks miles below Earth's surface feeding on nothing but traces of iron or potassium. However, these intrepid adventurers are rare and generally we see a greater number of species in areas near to the equator. Species diversity then reduces towards increasingly high latitudes. This pattern emerges because low-latitude environments are more stable and gain more solar energy, leading to more complex ecosystems. This is critical because, in their simplest form, plants depend on sunlight to grow and animals need plants to eat (or might feed on animals that eat plants), therefore more species can co-exist in environments nearer the equator. Traditional wisdom holds that tropical areas harbour more species because they are 'cradles of diversity' – areas where speciation is common. Contrary to this, recent scientific investigation has suggested that speciation might be more frequent in high-latitude environments, but we see fewer species because extinction is more common owing to the less benign conditions.

RELATED TOPICS
See also
MAKING SPECIES: ISOLATION
page 40

FROM ADAPTATION TO
SPECIATION
page 48

EVOLUTIONARY RATES &
EXTINCTIONS
page 86

30-SECOND TEXT
Chris Venditti

Warmer, sunnier regions near the equator are home to more various species than inhospitable higher latitudes. Two-thirds of all species live in rainforests.

NATURAL SELECTION

Chromosomes Extremely long molecules of nucleic acid and associated proteins found in cells. A chromosome carries genes and other sequences of base pairs, as well as attached proteins that regulate the action of the DNA.

Directional selection A form of natural selection in which particular physical characteristics are favoured by environmental conditions. The classic example is the variations of finch beak size, where availability of different sizes of seeds can results in large or small beaks dominating.

Disruptive selection Directional selection in two simultaneous opposite directions, disruptive selection occurs when an environment offers opportunities favouring two extremes of the possible phenotypes, resulting in a move from a mixed population to two populations exhibiting the extremes. Over time the two extreme variants can form separate species, like the varied species of finches observed by Charles Darwin on the Galapagos Islands.

DNA (Deoxyribonucleic acid) DNA is a long organic molecule that plays a vital role in the ability of living organisms to reproduce. The familiar double helix is a pair of polymers of bases whose sequence constitutes information, like the ones and zeroes of computer data.

Genetic diversity/gene pool In a particular species there will be a range of possible gene combinations – the genetic diversity refers to the number of different variants occurring in a population, which can be a measure of the species' ability to cope with changes in environment, since the greater the genetic diversity, the more chance there is of a variant helping the species to thrive in new conditions. Without genetic diversity there is little opportunity for evolution by natural selection, because there are no variants to select between. The gene pool is the whole collection of genes available within the population.

Genetic fingerprinting Also known as DNA testing or profiling, this is a set of techniques that uses specific sequences of DNA to identify individuals or biological relationships like that between parent and child. The process takes small segments of DNA and often looks for 'short tandem repeats', in which a pattern of bases is repeated a number of times, and the number of these repeats varies from individual to individual.

Genome The complete set of genetic information coded in DNA or RNA in a living organism. This includes both the genes, and the nucleic acid in a chromosome which does not code for production of proteins but still has important functions in the reproduction and functioning of cells. When a genome is 'sequenced', the order of base pairs in the DNA or RNA is recorded.

Industrial melanism The selection of genetic variants of organisms with darker pigments in an environment blackened by soot and other industrial pollutants. The best-known examples is moths that developed a darker colouring to make them harder to spot by predators on soot-darkened tree bark, meaning naturally darker moths were more likely to reproduce and pass on their dark pigmentation to their offspring.

Natural selection A prime mechanism of evolution, natural selection describes evolutionary change resulting from variation among individuals in a population, which leads to differences in reproductive success. Originally framed in comparison with artificial selection, in which traits were emphasized in animals and plants by selective breeding.

Phenotype The visible characteristics of a specific organism, sometimes used to contrast with the 'genotype', which is the total set of genetic instructions – the combination of all the genes in a cell or organism.

Population In an evolutionary context, the set of individuals of a species in a particular environment that have the potential to breed during their lifetimes and as a result can influence the evolution of the species within that habitat.

Stabilizing selection The evolutionary equivalent of the statistical 'regression to the mean', stabilizing selection is a tendency to favour phenotypes that are not at one extreme or the other, but are somewhere in between. In environments in which the extremes are less likely to survive to reproduce, the result is negative selection, favouring the 'average' individuals. Stabilizing selection decreases genetic diversity.

POPULATIONS

the 30-second evolution

The definition of species is
contentious, but a population is more
straightforward: a group of individuals in
one geographic area that can interbreed.
A species can be made up of a single population
– such as the Baikal seal, *Pusa sibirica*, the
world's only freshwater seal, confined to one
lake in Siberia – or many populations, such as
Alpine plants on the tops of separate
mountains. We can measure how much
difference exists between populations, and use
this to understand natural history – for example,
in the eyeless Mexican cave fish, *Astyanas
mexicanus*, populations found near the surface
are all very similar, but every deep cave hosts a
unique population, suggesting fish very rarely
find their way in or out of deep caves. Evolution
works differently in large and small populations.
Where there are many individuals, there are
more chances for a useful mutation to occur: as
an extreme example, in the virus population
inside one HIV-infected patient, every single
possible mutation occurs daily. Small
populations, on the other hand, lose genetic
diversity quickly, and chance plays a larger part
in their evolution. Crashes in population size –
such as those experienced by many endangered
species – leave populations at risk of inbreeding
depression or extinction.

RELATED TOPICS
See also
SPECIES & TAXONOMY
page 36

GENETIC VARIATION
page 64

3-SECOND THRASH
An individual organism
does not evolve; rather,
it is the population that
evolves and diverges over
the generations.

3-MINUTE THOUGHT
Sometimes, only a fraction
of a population has a
chance to breed. Consider
elephant seals where one
male holds a harem of 100
females. These populations
evolve as if they were
smaller than a count of the
individuals would predict:
that is, they have a low
effective population size.
An imbalanced ratio of
males to females will
reduce the effective
population size; so will
natural selection.

3-SECOND BIOGRAPHY
SEWALL WRIGHT
1889–1988
American geneticist who
helped found the field of
population genetics and the
'New synthesis of evolutionary
biology', linking genetics and
natural selection

30-SECOND TEXT
Louise Johnson

*Several populations
of the eyeless cave
fish have been found
in a single Mexican
cave, but the Baikal
seal lives only in
Lake Baikal, Siberia,
and has only one
population.*

THE NEED FOR ADAPTATION

the 30-second evolution

Life thrives in many seemingly hostile environments: parched deserts, freezing tundra and deep oceanic trenches at unimaginable pressure. Special features – adaptations – are needed to survive these extremes, such as the heat-resistant proteins found in bacteria from boiling hot springs, or the cellular antifreeze found in cold-climate plants to prevent ice crystals rupturing their cells. However, our own environment is also a hostile one, requiring as much specialist adaptation as any other. We breathe air, which is a feat only a few animal groups have accomplished, and even the presence of oxygen molecules is lethal to some microbes. Wherever organisms exist, there will be challenges to overcome, so they must evolve tricks and techniques to suit the world they inhabit, and as that world changes over time, they must adapt further to catch up. Prey, predators, diseases and competitors are important parts of the environment, and being themselves subject to evolution, they are likely to be a particularly rapidly changing part. This means that in one sense at least, the habitats that appear most hospitable, and in which the widest variety of species flourish, are the very hardest places to make a living. There is truly no such thing as an easy life.

3-SECOND THRASH
Living things evolve to suit their environments; this process, adaptation, is both a necessary and an inevitable one.

3-MINUTE THOUGHT
Luckily, adaptation is as unstoppable as it is vital. Where there is genetic variation that affects the ability to survive or reproduce, adaptation logically must occur. Complex adaptations may build up over thousands or millions of generations of selection, but if heritable variation exists – and it almost always does – natural selection is an inescapable force and greater adaptation results.

RELATED TOPIC
See also
GENETIC VARIATION
page 64

3-SECOND BIOGRAPHY
JEAN-BAPTISTE LAMARCK
1744–1829
French naturalist who is widely known for his theory of inheritance of acquired characteristics, better known as Lamarckism.

30-SECOND TEXT
Louise Johnson

Life adapts to hostile environs ... from the oryx in the deserts of Africa to the arctic hare in snowbound polar regions.

GENES

the 30-second evolution

Most biologists today would define a gene in molecular terms – a length of DNA that a cell uses to produce a protein – but in fact, the word was coined decades before DNA was known to play any role in heredity. It described the hypothetical cause of a difference between individuals: a short pea plant and a tall one, or a red-eyed and a white-eyed fly. Where there was no difference, there could be no gene. During the 20th century, scientists discovered that genes were associated with chromosomes, the then mysterious string-shaped objects seen in dividing cells, and that most genes corresponded to proteins. Slowly it became clear that genes were *information*. Genetic information is stored in DNA, a long, inert molecule made up of small subunits, and genetic differences are alterations or mistakes in this information. Such mistakes can alter the protein encoded by the gene – for example, the gene for cystic fibrosis makes a malfunctioning version of a protein that controls the salt content of bodily fluids. However, as we understand whole genomes better, the narrow molecular definition of the gene can turn out to be incomplete: some genes exert their influence through other means.

RELATED TOPICS
See also
THE MODERN SYNTHESIS
page 28

MUTATION & SPECIATION
page 46

3-SECOND BIOGRAPHIES

THOMAS HUNT MORGAN
1866–1945
American geneticist who won the Nobel Prize for his work on the role of chromosomes in physiology in 1933

LILIAN VAUGHAN MORGAN
1870–1952
American geneticist who helped popularize the use of *Drosophila melanogaster* as the model species for the study of genetics

ALFRED HENRY STURTEVANT
1891–1970
American geneticist who also worked on the effects of atomic bombs on human populations

30-SECOND TEXT
Louise Johnson

A pest in restaurants and at home, but a favourite among geneticists – the common fruit fly.

3-SECOND THRASH
We've all got heads, and genes cause heads to grow. So in the original sense of the word, nobody has genes for growing a head.

3-MINUTE THOUGHT
In 1911, Alfed Sturtevant, then an undergraduate student hired to wash up in T.H. Morgan's laboratory, made a breakthrough. It was known that some genes were 'coupled' – if you inherited one, you would probably inherit the other, too – and using experimental crosses, the strength of coupling could be measured exactly. Sturtevant realized that coupled genes were physically close together, and drew the first genetic map showing their positions relative to one another.

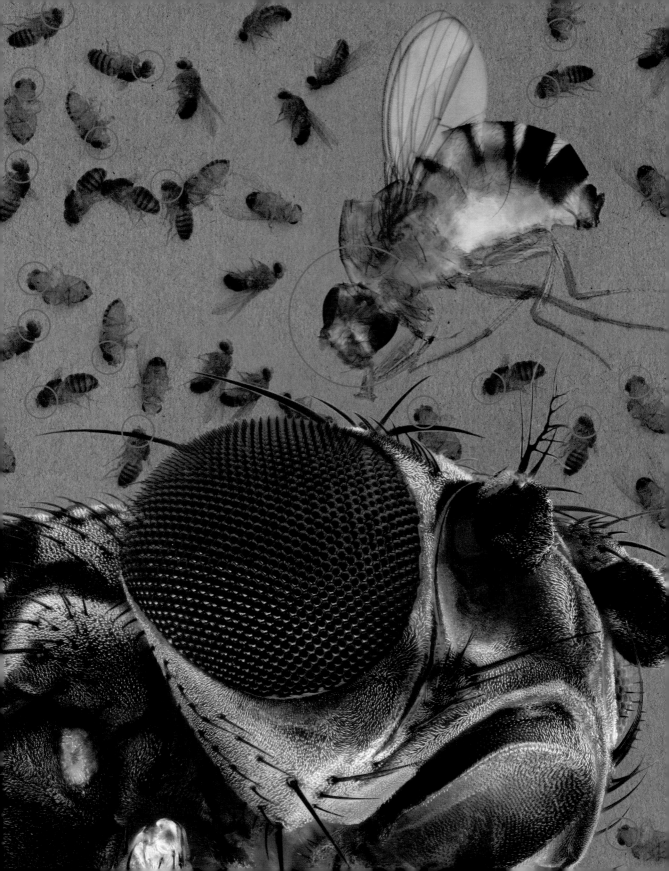

5 November 1892
Born in Oxford to John and Louisa Haldane

1911
Attends New College, Oxford

1914
Commissioned as second lieutenant in the First World War

1919
Made fellow of New College, Oxford, working in physiology and genetics

1922
Becomes a reader in biochemistry at Trinity College, Cambridge

1924-34
Publishes breakthrough series of papers 'A Mathematical Theory of Natural Selection'

1926
Marries Charlotte Burghes

1926
Publishes essay 'On Being the Right Size' which establishes 'Haldane's principle'

1929
Introduced the 'chemical soup' concept of the origin of life

1932
Elected Fellow of the Royal Society

1933
Appointed professor of genetics at University College, London

1945
Marries Helen Spurway

1956
Moves to Calcutta to join the Indian Statistical Institute

1961
Becomes Indian citizen

1 December 1964
Dies at Bhubaneswar in India

J.B.S. HALDANE

The geneticist and biometrician

John Burdon Sanderson Haldane (known in publications as J.B.S., but as Jack socially) had a surprising education for someone who would have such an influence. His father was a physiologist with his own home laboratory, where Haldane picked up an enthusiasm for science, yet on going to Oxford from Eton he studied mathematics and later converted his degree to Greats (the colloquial shorthand for humanities, classics and divinity). After the First World War, in which he served in France and Iraq, Haldane returned to Oxford as a fellow working in genetics and physiology – remarkable considering he had no education in these subjects. He moved on, via ten years at Cambridge, where he did original work on enzyme kinetics, to spend most of his academic career at the University of London.

Haldane's greatest contribution to genetics came in his series of ten papers 'A Mathematical Theory of Natural Selection', which were pulled together in *The Causes of Evolution*. His development of a mathematical basis for natural selection became a central element of the 'new synthesis' that brought together Darwinian natural selection and Mendelian genetics. It was rare for a biologist of the day to have much mathematical

expertise, and Haldane made great use of his, bringing quantitative skill to play across a range of topics. A good example was his essay 'On Being the Right Size', which included 'Haldane's principle' that the size of a living organism puts restrictions on the bodily structures and mechanisms it can employ.

In his early 60s, Haldane and his new wife, Helen Spurway, emigrated to India, where he would live for the rest of his life. He claimed this was a response to the Suez crisis, though it also seems to have been connected to his wife's arrest for drunken misconduct, which led to her dismissal from London University and Haldane's subsequent withdrawal from his professorship. In India, Haldane worked at statistical and biometry units in Calcutta and Odisha, becoming an Indian citizen. He continued to work into the 1960s and in 1963 made an early use of the term 'clone' in reference to humans.

Haldane's wit was famous. Referring to genetic altruism, he once said 'I'd lay down my life for two brothers or eight cousins', while in a speech about life on other planets he commented that if there is a Creator, he clearly has a special preference for beetles, as there are 400,000 species of beetles on Earth but only 8,000 species of mammals.

GENETIC VARIATION

the 30-second evolution

People vary in many ways: some are taller than others, have different eye colours or wear different colours of nail varnish. Some of this is due to differences in genes: eye colour differences are almost entirely genetic, and height is partly genetic (nutrition and disease also have an influence on how tall a child grows). Natural selection can only lead to evolution if the differences it selects are caused by genes, so selection on eye colour would work more quickly than equally strong selection on height. Selection on nail varnish colour would be very ineffective, as this chosen characteristic is not determined by genes but largely by cosmetics manufacturers; its heritability will be close to zero. Heritability is a number describing the fraction of variation in a chosen characteristic, such as height, that is genetic rather than caused by differences in environment. It applies to a particular population in a particular environment – it is not a 'nature versus nurture' comparison, and it predicts very little about what might happen if the environment changed. If a headline reads, for instance, '40 per cent of drug addiction is due to genes', it's probably reporting a measure of heritability – and it's almost certainly wrong.

3-SECOND THRASH
Natural selection works on genes, but genes are not the only things that make us different from one another.

3-MINUTE THOUGHT
The heritability of human characteristics is often measured by comparing twins. The logic behind this is that a pair of non-identical twins will share their upbringing, their environment and half their genes, whereas identical twins share upbringing, environment and all their genes. If identical twins tend to be more similar in height than non-identical twins, this is evidence for genetic variation in height.

RELATED TOPIC
See also
POLYMORPHISM &
GENETIC DRIFT
page 66

3-SECOND BIOGRAPHIES
FRANCIS GALTON
1822–1911
British statistician who developed the concept of eugenics and studied the role of variation in human populations

DOUGLAS FALCONER
1913–2004
British geneticist renowned for his contribution to quantitative genetics, the use of mathematics to predict how traits will be inherited

30-SECOND TEXT
Louise Johnson

Identical twins are a valuable resource for those studying heritability. They share 100 per cent of their genes.

POLYMORPHISM & GENETIC DRIFT

the 30-second evolution

RELATED TOPICS
See also
GENETIC VARIATION
page 64

INDUSTRIAL MELANISM
Page 104

3-SECOND BIOGRAPHY
BERNARD KETTLEWELL
1907–79
English zoologist who
researched effect of industrial
melanism on moths

30-SECOND TEXT
Louise Johnson

3-SECOND THRASH
Humans, like most species,
are highly polymorphic:
you are unique, and so is
everybody else.

3-MINUTE THOUGHT
With today's technology
we can reveal even more
polymorphism than a
genetic fingerprint: we can
tell identical twins apart,
or pinpoint the source of a
tumour that has spread
through a patient's body.
Each of us has 60 new
mutations never seen
before; however, nearly all
of these have no effect
whatsoever on either
appearance or behaviour.

Evolutionary biology has shown
us not only how species vary over time, but also
how much the members of one species vary
from each other. The idea that one 'wild type'
prevailed, from which there were only rare
departures, was challenged by studies of
animals that came in conspicuously different
forms – polymorphic species. Sometimes
there were clear reasons for a particular
polymorphism, such as in peppered moths,
whose light and dark forms were well
camouflaged on lichen-covered and soot-
stained trees respectively. Often, though, the
causes of polymorphism were less obvious, such
as the banded snail (*Cepaea* spp.) whose diverse
colour and stripe patterns are still mysterious
– they may provide camouflage, confuse
predators or both. When molecular biology
techniques began to be widely used, they
revealed levels of polymorphism far higher than
expected: genetic fingerprinting was soon able
to distinguish between any two individuals
(except clones or twins) of any species (except a
few highly inbred species, such as cheetahs).
Polymorphism is the rule, not the exception.

*Identical twins have
unique fingerprints.
Scientists do not
know what benefits
arise from some
polymorphic forms –
like those of
banded snails.*

TYPES OF SELECTION

the 30-second evolution

A diagram frequently used to illustrate natural selection is that of a giraffe-like animal straining successfully to reach the leaves of a tall tree while several shorter-necked specimens look on in hunger and envy. This is an example of directional selection, the most familiar type: one extreme is favoured, so selection results in a straightforward change to the population – in this case, longer necks. A different type of selection is stabilizing selection, in which what is favoured is neither one extreme nor the other, but a happy medium. In humans, stabilizing selection may act on birth weight, as unusually small or large babies have poorer health outcomes. Stabilizing selection causes a subtler change than directional, in that the population merely becomes 'more average' rather than the average itself changing, but it is likely to be very common in the wild: a frog protected by camouflage will live longest if it is neither too dark nor too light green in colour. Disruptive selection, in which both extremes are favoured, is rare but has the interesting effect of increasing the variation in a population, or even splitting it into two divergent groups.

RELATED TOPIC
See also
UNITS OF SELECTION
page 70

3-SECOND BIOGRAPHY
CHARLES DARWIN
1809–82
English naturalist and geologist who first described a form of selection in On the Origins of Species

30-SECOND TEXT
Louise Johnson

3-SECOND THRASH
The survival of the fittest can in some cases mean the survival of the most average.

3-MINUTE THOUGHT
Two or even all three types of selection can in theory happen at the same time to the same population, if they work on different features of the animal or plant in question. The same is true of artificial selection: a breeder might aim to create apple trees with the highest possible yield (directional selection), but consistent fruit size (stabilizing selection).

In giraffes selection may favour a longer neck, for reaching leaves on tall trees; in frogs it may favour middling rather than extreme greens, to make it easier to hide.

UNITS OF SELECTION

the 30-second evolution

3-SECOND THRASH
Genes build bees to make more genes. Bees build hives to make more bees. All three are reproducing – but who's evolving?

3-MINUTE THOUGHT
For single-celled life (which is, after all, most life) the individual and the cell are the same unit. Animals, on the other hand, are a co-operating collective of cells: if cells begin to replicate independently, the result is cancer. As a result, animals have evolved to reduce the opportunity for cell-level evolution. This may explain why our body cells have built-in limits to how many times they can divide.

Genes replicate, cells divide, individuals breed, groups propagate themselves, and species speciate: the life we see around us has been successful in all these ways. But on what level – genes, cells, individuals, groups or species – does natural selection act? Most biologists would single out the gene as the unit of selection. Genes are the basis of heredity for all the other units, and copy themselves more exactly. Adaptations that benefit an individual also promote the spread of its genes, and adaptations that benefit groups, such as the waggle dance used by a honeybee to direct her sisters to nectar, are well explained by gene-level selection: the bee who helps her sister helps the genes that they both share. Taking a gene's eye view explains a lot of otherwise puzzling features: for example, the fact that although one male is needed to fertilise many females, most sexual species appear to waste fully half of their resources on making males. Many genes also reproduce themselves without doing anything for their host, but simply by cutting, copying or pasting themselves around the chromosomes. These jumping genes make up about half of human DNA: 50 times more than is needed to build our bodies.

RELATED TOPICS
See also
THE NEED FOR ADAPTATION
page 58

GENES
page 60

3-SECOND BIOGRAPHIES
WILLIAM D. HAMILTON
1936-2000
British evolutionary biologist, whose found fame through his theoretical work on kin selection and altruism

BARBARA MCCLINTOCK
1902-1992
American geneticist and the 1983 Nobel laureate in physiology or medicine

30-SECOND TEXT
Louise Johnson

The honeybee's waggle dance – which shows other bees the way to nectar – is made possible by an adaptation that benefits the group.

EVOLUTIONARY HISTORY & EXTINCTION

Acritarchs A catch-all term for small fossils of organic structures found in sedimentary rocks that do not match any other classification (the word means 'doubtful origin'). Acritarchs date back as much as 3 billion years, but peak in the Palaeozoic era, 540 to 250 million years ago.

Brachiopods Literally an 'arm-foot', a phylum or division of marine invertebrates sometimes called 'lamp shells' because of the resemblance of some brachiopods to classical oil lamps. A brachiopod's two hinged shells form a top and bottom, whereas a bivalve like a mussel or clam has shells enclosing its two sides. Most brachiopods attach themselves to a surface with a fleshy anchor called a pedicle.

Cambrian The earliest part of the Palaeozoic era, the geological period from about 540 to 485 million years ago when most modern animal types appeared. A particularly extreme diversification of marine forms between 540 and 525 million years ago is known as the Cambrian explosion.

DNA (Deoxyribonucleic acid) DNA is a long organic molecule that plays a vital role in the ability of living organisms to reproduce. The familiar double helix is a pair of polymers of bases whose sequence constitutes information, like the ones and zeroes of computer data.

Endosymbiosis 'Endo' simply means 'within', making endosymbiosis a symbiotic relationship where one organism is contained within the other. It is thought that the mitochondria that act as power sources for eukaryotic cells originated as bacteria that had an endosymbiotic relationship with another single-celled organism.

Eukaryotic cells Cells with a nucleus, an inner structure with a surrounding membrane that contains most of the cell's genetic material. All organisms with more than one cell, including animals, plants and fungi, have eukaryotic cells. The alternative, prokaryotic cells, typical of bacteria and archaea, have no nucleus. 'Eukaryote' translates literally as 'good kernel', where prokaryote is 'before kernel'.

Gaia hypothesis The ancient Greek word for Earth, Gaia was the goddess who personified the planet. The term was used by English environmentalist James Lovelock to describe a hypothesis that the global ecosystem can be considered a single, self-regulating organism in which organisms evolve in a feedback mechanism with their environment.

Mass extinction A period in which a large percentage of life on Earth was wiped out, resulting in a significant drop in diversity and the opportunity for new species to occupy niches previously dominated by others. Also known as an extinction event.

Molecular clock The idea that it is possible to estimate the relative amounts of time since species diverged from a common ancestor by changes in DNA sequences in which the changes appear to have taken place at a rate that is constant over time.

Monotreme fauna or monotremes An unusual order of mammals that only have a single excretory and reproductive orifice (hence monotreme or 'single hole') and lay eggs. Once more widespread than they are today, the only remaining living examples are the platypus and echidnas of Australia.

Organelle An internal part of a eukaryotic cell, usually surrounded by its own membrane. Includes mitochondria and plastids. Sometimes also applied to 'molecular machines' without a membrane, such as the ribosomes that translate messenger RNA into proteins.

Punctuated equilibrium An evolutionary theory championed by American biologist Stephen Jay Gould, according to which species exist for long periods of time with very little evolutionary change, punctuated by relatively rapid events in which a species splits into two individual species. The most common alternative, phyletic gradualism, assumes that gradual change eventually results in variants becoming distinct species.

RNA (Ribonucleic acid) Messenger RNA is produced from gene-encoding regions of DNA and translated into protein on ribosomes (composed of ribosomal RNA), using transfer RNAs that deliver the amino acids needed to construct the protein. Some viruses have genomes made up of RNA rather than DNA.

Speciation The formation of new species, often as a result of change in environment that favours a particular genetic variant, though also contributed to by genetic drift.

Species drift A change in the frequency of a particular variant of a gene that is not due to a selective process but rather to random fluctuations and results in speciation.

HOW LIFE BEGAN

the 30-second evolution

Philosophers may ponder why we exist, but for biologists the real question is how chemicals became life. Strange as it seems, it is possible that the ingredients of life fell from the stars. Life is an ongoing chemical reaction that is carried out by specialized machinery made of components that can be found in comets and meteorites – or anywhere else that space dust and water encounter an energy source like sunlight. Think of every living cell as computer hardware that runs some version of an operating system that we will call Life 2.0. But what was Life 1.0, and how did it work? At the heart of our living computer is a processor that reads instructions written on long strings of RNA and makes proteins. The processor is made of chains of RNA, and burns single RNA molecules as a power source; it is one of many RNA-built, RNA-fuelled machines that make up our living computer. These discoveries inspired scientists such as Walter Gilbert to imagine a time when chemicals began to change from inert to alive, and the future of all living things floated in a pool of ever-evolving, self-copying RNA. Thus Gilbert proposed that our world arose from a more ancient RNA world.

RELATED TOPICS
See also
TRANSMUTATION & ARCHETYPES
page 18

VARIATION & SELECTION
page 20

ANCESTORS & TIMESCALES
page 138

3-SECOND BIOGRAPHIES
STANLEY MILLER
1930–2007
American biochemist who showed how the molecules of life can come together from simple chemicals

WALTER GILBERT
1932–
American biologist and physicist who proposed and named the RNA world

30-SECOND TEXT
Ben Neuman

3-SECOND THRASH
From the heavens may have come RNA: quite possibly the true architect of the machinery of life.

3-MINUTE THOUGHT
Some clues about life in the RNA world come from another strange group of self-copying RNA strands – we know them as the viruses that cause the common cold and AIDS. Every living thing runs on similar biological hardware except for these RNA viruses, which may be the last vestiges of the RNA world. These viruses dwell among us as aliens, not from outer space, but from our distant past.

Are we the product of space dust, water and sunlight? Walter Gilbert's hypothesis is that life on Earth was generated in a pool of self-replicating RNA.

THE GEOLOGICAL RECORD

the 30-second evolution

3-SECOND THRASH
Tiny bacterial belches became the gale-force winds of evolutionary change that have shaped every facet of the living world around us.

3-MINUTE THOUGHT
What else can the acritarchs tell us? Most are biological mysteries that we struggle to place on the tree of life. What we do know is that about 1 billion years ago, acritarchs began to develop the sort of spiny armaments that would make life difficult for multicellular predators. They have diversified whenever larger animals have done so, and suffered in all the major extinctions. Whatever they are, our fates have always been linked.

Earth's history is written in stone
– the rocks beneath our feet contain layer upon layer of geological narrative. These layers are like a scrapbook filled with mementos that stretch back 500,000 times further than recorded human history. The earliest pages tell of a period of a billion years when our planet was a red-hot ball of liquid metal, devoid of breathable air and utterly inhospitable to life. The turning point came about 3.5 billion years ago with the arrival of oxygen – a form of gaseous waste product puffed out by photosynthetic bacteria. The oxygen reacted with silicon, phosphorus and calcium to make new chemicals that could dissolve in water. Occasionally the oxygen-containing chemicals became so concentrated that they made a crystalline film over rocks and even living things. These early organisms – called acritarchs – became the first fossils. Acritarch fossils probably show evolution's early steps, but they are remarkably difficult to interpret. Oxygen can burn molecules the way a fire burns tinder, so early creatures were forced to adapt, hide or die as the air grew thick with this toxic gas. It may be the rise in oxygen levels that caused a few cells to team up, one living inside another. And these unlikely survivors, 2-billion-year-old cell-in-cell creatures, eventually became plants, mushrooms and even people.

RELATED TOPICS
See also
TRANSMUTATION & ARCHETYPES
page 18

EVOLUTIONARY RATES & EXTINCTIONS
page 86

ANCESTORS & TIMESCALES
page 138

3-SECOND BIOGRAPHIES
ARTHUR HOLMES
1890–1965
British geologist who worked out how to date rocks

LYNN MARGULIS
1938–2011
American biologist who proposed the theory of cells-in-cells

30-SECOND TEXT
Ben Neuman

Earth's most ancient history can be traced through acritarch fossils in sedimentary rocks; the name comes from Ancient Greek for 'uncertain origins'.

GEOLOGICAL CHANGE & MAMMALIAN EVOLUTION

the 30-second evolution

Time and drift are problems

common to both evolution and geology. Changes in species and changes in rocks can take a long time, which makes both difficult to study. And just as the drift of continents floating across Earth's molten metal core can make it difficult to interpret where and when geological events happened, our understanding of evolution can be confused by environmental change and the constant drift of species into new habitats. Fossils can help us understand evolution, but it turns out that nature has also left behind some living time-capsules – places that preserve forms of life that were once dominant, but are now virtually unknown in the outside world. One such time-capsule can be found in the islands connecting Australia to Asia, where biologist Alfred Russel Wallace noticed a gradual progression from the more primitive forms of Australian animals to the more modern Asian forms. Australia preserves strange mammals like the egg-laying platypus, which has ten sex-determining chromosomes to our X and Y, and the honey possum, which gives birth to the smallest offspring of any mammal. Australian mammals show that while some features like hair and milk production have changed little in the past 100 million years, there has been a revolution in the way we rear our young.

RELATED TOPICS
See also
MAKING SPECIES: ISOLATION
page 40

MECHANISMS OF ISOLATION
page 44

SPECIES DIVERSITY
page 50

3-SECOND BIOGRAPHY
ALFRED RUSSEL WALLACE
1823–1913
English evolutionary theorist and father of biogeography

30-SECOND TEXT
Ben Neuman

3-SECOND THRASH
While mammalian evolution was underway, tectonic processes led to Australia's isolation from Asia, resulting in the unique Australian marsupial and monotreme fauna.

3-MINUTE THOUGHT
Why did most mammals stop laying eggs? The credit or blame probably goes to a virus – a distant relative of HIV that has infected mammals so thoroughly that there are now thousands of old broken copies of the virus littering your DNA. Part of one virus, called syncytin, still works. Syncytin effectively transforms the pregnant mother into a walking egg by passing nutrition and oxygen across the womb.

Strange mammals include the tiny honey possum – which weighs half as much as a mouse – and the platypus, one of five mammal species that lay eggs.

5 March 1938
Born in Chicago to Morris and Leona Alexander

1952
Attends the University of Chicago Laboratory Schools

1957
Marries astronomer and science popularizer Carl Sagan

1960
Receives Masters in Biological Sciences from the University of Wisconsin-Madison

1965
Receives PhD from the University of Berkeley, California

1967
First suggested that organelles in eukaryotic cells were originally independent bacteria

1967
Marries crystallographer Thomas Margulis

1970
Publication of her groundbreaking book *Origin of Eukaryotic Cells*

1978
Paper by Schwartz and Dayhoff confirms Margulis's idea of endosymbiosis

1983
Elected to the US National Academy of Sciences

1988
Moves to the University of Massachusetts, Amherst

1999
Receives the National Medal of Science from US President Bill Clinton

2008
Receives the Darwin-Wallace medal

22 November 2011
Dies in Amherst, Massachusetts

LYNN MARGULIS

Lynn Petra Margulis's biggest
claim to fame derives from early in her career, at
age 29, just two years after receiving her PhD. It
was then that she put forward the idea of
endosymbiosis, suggesting that some early cells
incorporated bacteria, producing a more
complex structure in which the bacteria became
'organelles', enabling the cells to gain energy by
photosynthesis in the case of plants, or by
processing oxygen in animals and plants.

Margulis's idea was originally dismissed,
and her first paper rejected many times before
being accepted by the *Journal of Theoretical
Biology*. In part this was because prevailing
evolutionary theory was focused on random
mutations. Margulis challenged what she called
the 'ultra-Darwinian orthodoxy', arguing that
symbiosis, in which organisms that are mutually
beneficial may come together to form a single
organism, was more important at the level of
microorganisms and hence in the first 3 billion
years of life before complex organisms evolved.

While in Chicago, where she was an
undergraduate, she met Carl Sagan, her first
husband and one of two famous mavericks
of science with whom she was associated, the
other being James Lovelock, with whom
she controversially collaborated on the Gaia
hypothesis in the 1970s. This proposed that
the Earth acts in a self-regulating manner, with
geology, meteorology and life forms acting
together to allow the whole to perpetuate itself,
even though changes may be to the
detriment of individual species. In effect this
is also a form of symbiosis between the living
parts of Gaia and the environment, though
Margulis was quick to emphasize that Earth
is not a true organism, because it is not distinct
from its waste.

Margulis waited more than ten years for
experimental evidence to support her
breakthrough theory on the symbiotic origin
of complex cells, when a paper submitted to
Science magazine by Robert M. Schwartz and
Margaret O. Dayhoff, 'Origins of prokaryotes,
eukaryotes, mitochondria and chloroplasts',
established that chloroplasts shared a common
and recent ancestor with blue-green algae, and
mitochondria had a common ancestor with
bacteria known as *Rhodospirillaceae*.

Some of Margulis's later ideas never made
the mainstream. She suggested that
symbiotic relationships were the main way in
which genetic variation was achieved – by DNA
transfer between cells; doubted that HIV was
an infectious virus and the cause of AIDS;
and supported a suggestion that the larvae
and adults of metamorphosing species had
not evolved from the same ancestors. All
were regarded as fringe concepts yet this
great challenger of the biological status quo
was certainly successful in her first idea
about the bacterial origin of chloroplasts
and mitochondria.

Brian Clegg

EMERGENCE OF MAJOR PLANT & ANIMAL GROUPS

the 30-second evolution

3-SECOND THRASH
The first jellyfish were wise to frolic under Cambrian algal towers because 300 million years is far too long to wait for flowers.

3-MINUTE THOUGHT
A lot of luck goes into each fossil. First, there is bad luck for the subject of the fossil, then good luck with the arrival of minerals and bacteria in time to preserve it. Creatures that are already part mineral when they are alive stand the best chance of becoming fossils. For example, bones are a mineral called apatite and many plants store up silicon, both of which could lead to a place in the fossil record.

When did the familiar groups of plants and animals first appear? Fossils and genes combine to tell the story. Most animal groups appeared in the Cambrian. Among the earliest animal fossils were comb jellyfish – gelatinous, iridescent creatures that pulse through the ocean depths today. Remarkable three-dimensional fossils of 540-million-year-old baby comb jellies have been found in China, and when a comb jelly genome was finally sequenced in 2014 it showed that their genes may have originated before the genes found in more familiar animals. Even the organs that work like nerves and muscles in comb jellies seem to have evolved independently from those in the more familiar animals. The land plants we recognize today are of much more recent origin than Cambrian sea-creatures. The first moss-like land plants probably evolved from algae, the tiny light-loving organisms that form a greenish haze in ponds and fish-tanks. Algae are difficult to recognize as fossils because they vary in shape from sheets of sea lettuce to mighty kelp forests. The first recognizable land plant fossils date from about 430 million years ago, and it was only some 240 million years ago that Earth got its first fossil flowers.

RELATED TOPICS
See also
THE GEOLOGICAL RECORD
page 78

GREAT EXTINCTIONS
page 90

CAUSES OF EXTINCTION
page 92

30-SECOND TEXT
Ben Neuman

Comb jellyfish have been around for more than 500 million years and their genes may derive from a separate evolutionary history.

EVOLUTIONARY RATES & EXTINCTIONS

the 30-second evolution

Tick tock ... tick tock ... the traditional Darwinian view of evolution states that morphological and genetic divergence accumulates steadily as a function of time, with species arising and going extinct gradually. This is at the heart of the idea of the molecular clock – over the course of many millions of years, genetic mutations build up at a reliable rate on a given stretch of DNA. This is a potentially powerful concept that can be used to date divergence times between species or groups of species. While this gradual view of evolution remains persuasive, it is becoming supplanted by a more episodic perspective. This change started in the early 1970s when Eldredge and Gould proposed the controversial theory of 'punctuated equilibrium', which suggested that most evolutionary change is concentrated at the time of speciation events. Their theory was based on morphological evidence from the fossil record, but more recently punctuational change has been detected at a genetic level. Further doubt has also been cast on the idea that extinction is a gradual process through the identification of 'mass extinction' events, some which saw the demise of more than 90 per cent of species. Such events cause a major restructuring of life on Earth in what is, relatively speaking, an instant of geological time.

RELATED TOPICS
See also
FROM ADAPTATION TO SPECIATION
page 48

EMERGENCE OF MAJOR PLANT & ANIMAL GROUPS
page 84

GREAT EXTINCTIONS
page 90

CAUSES OF EXTINCTION
page 92

3-SECOND BIOGRAPHIES
STEPHEN JAY GOULD & NILES ELDREDGE
1941–2002 & 1943–
American palaeontologists and evolutionary biologists who proposed the theory of punctuated equilibrium

30-SECOND TEXT
Chris Venditti

3-SECOND THRASH
The view of evolution as a gradual process is attractive, but should we expect it to be this way on a dynamic planet such as Earth?

3-MINUTE THOUGHT
While it is often argued that evolution is either gradual or sporadic, the more likely explanation is that it is and has been both. Species encounter extremely changing environments and this is especially true over the course of geological time. Since life on Earth evolved our planet has encountered enormous fluctuations in temperature and land structure (to name just two of many things). We must surely expect these factors to have influenced evolutionary rates.

Time runs on, but are evolution and extinction gradual or more episodic?

THE MYSTERY OF THE CAMBRIAN EXPLOSION

the 30-second evolution

Look at rocks laid down about half a billion years ago: they contain fossils that look like miniature versions of modern animals. You may notice some changes – the first squid only had two tentacles, and early relatives of spiders had forked heads and over a dozen legs – but modern-looking body plans were in place. But how did they get there? Look back a further 20 million years and life was almost unrecognizably weird. Familiar creatures like jellyfish shared the sea with metre-wide ribbed ovals like *Dickinsonia rex*, and stretchy cucumber-shaped *Kimberella* that scratched feeding trails wherever they went. Oddly, most of these primordial sea-dwellers lacked anything resembling a head. What happened next is a mysterious time called the Cambrian explosion. We know the Earth cycled rapidly from ice age to global warming and back again: did animals that evolved primitive heads succeed in the Cambrian after extreme climate change wiped out the competition? Unfortunately the fossil record is patchy here. However it happened, over the course of 20 million years, the headless oddities of old gradually gave way to a rather modern-looking variety of body plans.

RELATED TOPICS
See also
EVOLUTIONARY RATES & EXTINCTIONS
page 86

SEX & EVOLUTIONARY ARMS RACE
page 130

3-SECOND BIOGRAPHIES
CHARLES DARWIN
1809–82
British naturalist and geologist who provided a compelling explanation of why we see such an explosion in diversity during the Cambrian

LYNN MARGULIS
1938–2011
American evolutionary biologist who came up with an elegant explanation of how life arrived in the Cambrian explosion

30-SECOND TEXT
Ben Neuman

3-SECOND THRASH
Could a Cambrian climate catastrophe be the break our ancient relatives needed to get ahead?

3-MINUTE THOUGHT
Evolutionary changes can seem sudden, but the Cambrian explosion was probably more of a slow metamorphosis than a biological big bang. Little changes as embryos grow can lead to huge differences in the adult. For example, genes like collagen and laminin that form a skin around the simplest single-celled animals are the same ones that hold multicellular animals together. Animals are all working with the same genetic tools – some just use them differently.

The oval Dickinsonia rex *was one of many weird and wonderful pre-Cambrian sea-dwellers.*

GREAT EXTINCTIONS
the 30-second evolution

3-SECOND THRASH
Life on Earth barely
escaped the Medusa's
gaze, as a wave of
Permian CO2 became a
petrifying calcite crust.

3-MINUTE THOUGHT
Why did so many die?
Many sea creatures had
evolved stone-like
defences by capturing
calcium carbonate, just as
modern oysters make their
shells. Tropical reefs grew
so fast you could tell a
coral's age by counting
daily growth rings. But as
the seas' chemical
composition changed,
calcium carbonate
overwhelmed these
creatures. Throughout
the oceans, sturdy homes
became inescapable tombs,
as the bottom of the food
chain effectively turned
to stone.

Fossils are often a better record
of how things died than how they lived. A giant
asteroid probably hastened the dinosaurs'
demise 65 million years ago. But a greater, more
mysterious, extinction happened some 200
million years earlier, at the end of the Permian.
In the Permian, many-legged trilobites still
burrowed along seafloors filled with crinoids –
animals that resembled starfish on stems.
No-one is sure what happened next, but
Permian rocks tell us that a massive amount
of carbon dioxide dissolved in the ocean.
There, it combined with calcium until the water
contained so much calcium carbonate that
it began to crystallize out like salt does on the
shores of the Dead Sea. Today, entire mountains
of Permian calcium carbonate can be seen
around the world. Nearly nine-tenths of all
sea creatures died out, and with so many
ecosystems destroyed, it is a wonder that
anything survived. But as at the beginning
of the Cambrian, a brush with catastrophe
cleared the way for a new explosion of life. The
survivors became dinosaurs, modern insects and
the first mammals. Now there are indications
that climate change and CO2 are returning Earth
towards end-Permian conditions. If there is a
lesson from the Permian, it is that once begun,
great extinctions are hard to stop.

RELATED TOPICS
See also
THE GEOLOGICAL RECORD
page 78

THE MYSTERY OF THE
CAMBRIAN EXPLOSION
page 88

CAUSES OF EXTINCTION
page 92

30-SECOND TEXT
Ben Neuman

*Calcium carbonate,
CaCO$_3$, drove the great
extinction at the end
of the Permian period.
Is climate change
driving us back to
these conditions?*

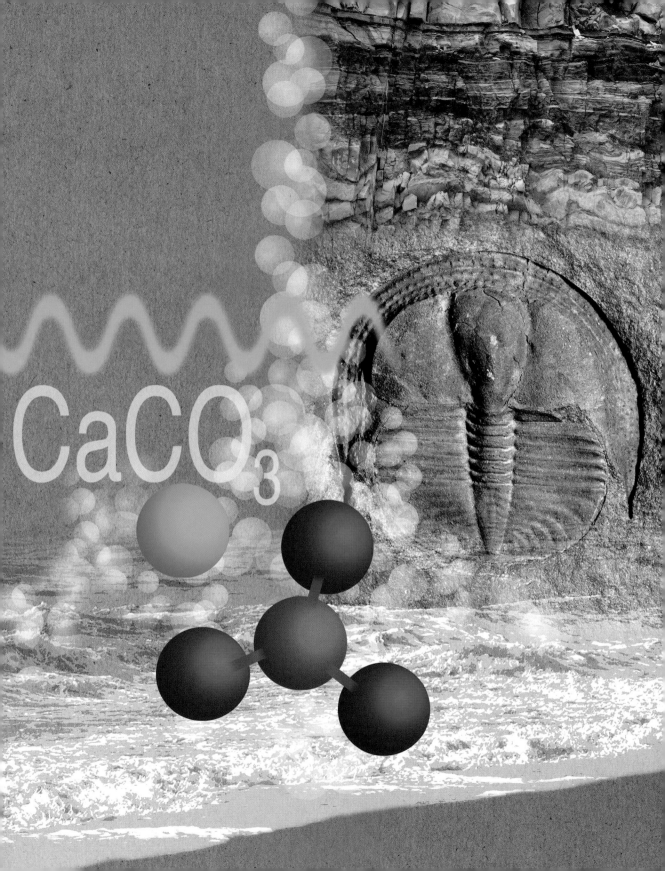

CaCO₃

CAUSES OF EXTINCTION

the 30-second evolution

Darwin noted that extinction happens gradually. A species dwindles, until competition or some unlucky event finally nudges it over the edge. Take brachiopods, grim survivors of 540 million years of undersea extinctions. They resemble clams, but grow slowly and are adapted to withstand starvation. Brachiopods nearly became extinct when the seas filled with methane and calcium carbonate at the end of the Permian. Did the surviving brachiopods lose too much genetic diversity or did slow metabolisms make them uncompetitive? Whatever the cause, clams bounced back; brachiopods did not. Fifty million years later, an end-Triassic extinction driven by climate change and ocean acidification reduced brachiopod diversity again. Finally, an impact from outer space ended the dinosaur dynasty and left only a few isolated brachiopods. From their peak as the dominant fossil-formers to arise from the Cambrian Explosion, to their present existence clinging to life only where the ocean is too deep or cold for clams to thrive, brachiopods have narrowly escaped extinction again and again. Now that humans can colonize every environment from seabed to space, we have ourselves become a major cause of extinction; come the next big extinction, will we die like dinosaurs or brazen it out like brachiopods?

RELATED TOPICS
See also
GEOLOGICAL CHANGE & MAMMALIAN EVOLUTION
page 80

EVOLUTIONARY RATES & EXTINCTIONS
page 86

THE MYSTERY OF THE CAMBRIAN EXPLOSION
page 88

3-SECOND BIOGRAPHIES
CHARLES DARWIN
1809–82
English naturalist and geologist who studied the causes of extinction

PAUL CRUTZEN
1933–
Dutch atmospheric chemist who popularized the term 'Anthropocene'

30-SECOND TEXT
Ben Neuman

3-SECOND THRASH
The lesson of the brachiopods is that extinctions don't end life's competition – they just change the players and some of the rules.

3-MINUTE THOUGHT
All life forms are adapted to specific environments, but we humans differ in our ability to bend nature into a more human-friendly shape. As we remake the world in our own image, we aid the creatures that thrive around people – such as rats, cockroaches, racoons and ragworts. What lies ahead in the Anthropocene (the age of humanity)? Will we dwindle and expire or will technology be the bridge that carries us across the chasm of extinction? Time will tell.

The scales of life weigh brachiopods up against clams. So far, both have survived. Will humans?

EVOLUTION IN PROGRESS

Adaptive radiation The rapid production of new species from a single original species, particularly occurring when a significant change in the environment opens up new ecological niches.

Allopatric speciation A form of speciation in which populations of the same species are isolated from each other by a significant change in the habitat, leaving them to evolve separately without the opportunity to interbreed with different environmental pressures.

Escape and radiation coevolution A mechanism whereby a species undergoes a genetic change that gives it a new defence from a predator or environmental pressure. The reduction in selection means that the species is able to undergo rapid speciation, making use of ecological niches previously unavailable to it.

Ethology The study of animal behaviour, particularly in the natural environment.

Eusociality A type of animal behaviour involving subgroups with distinct roles, and a shared brood, usually produced by a single 'queen'. The subgroups often lose the abilities required for other roles. Mostly ants, bees, wasps and termites, though mole rats are eusocial mammals. 'Eusocial' literally means 'good social' implying that it is the most organized of the social forms.

Genetic diversity/gene pool In a particular species there will be a range of possible gene combinations – the genetic diversity refers to the number of different variants occurring in a population, which can be a measure of the species' ability to cope with changes in environment, because the greater the genetic diversity, the more chance there is of a variant helping the species to thrive in new conditions. Without genetic diversity there is little opportunity for evolution by natural selection, because there are no variants to select between. The gene pool is the whole collection of genes available within the population.

Imprinting A kind of learning behaviour that happens at a specific age or stage of animal development. The best-known examples are the way in which human children recognize their parents, and the way birds latch onto an early moving stimulus, even if it isn't a bird.

Kin selection The idea that evolutionary processes can work to the benefit of related organisms even if they are to the detriment of an individual, hence explaining some altruistic behaviour in which an individual makes a sacrifice in order to benefit its kin. Most clear in eusocial organisms where some individuals lose the ability to reproduce.

Mutual influence The interlinked development of two evolutionary processes. So, for example, the evolution of a species could have an impact on the environment, and the changes to the environment could have an impact on the species.

Natural selection A prime mechanism of evolution, natural selection is a tendency for particular characteristics to make it more likely for an individual to reproduce successfully in a particular environment, resulting in those characteristics becoming more common in the population. As a corollary, other characteristics will become less common. Originally framed in comparison with artificial selection, in which traits were emphasised in animals and plants by selective breeding.

Phylogeny/phylogenetic tree A branching diagram, sometimes called a 'tree of life' that shows evolutionary links between species. These were originally based on physical characteristics, but are now more likely to be dependent on genetic similarity. Phyla (sing. phylum) are a taxonomic division below kingdom and above class.

Symbiotic relationship When two species have a close interaction with each other it is termed symbiosis (literally 'living together'). Originally taken to refer purely to mutually beneficial interaction, it now includes less symmetrical relationships such as parasitic symbiosis.

Sympatric speciation Unlike allopatric speciation, here new species are formed from a single original species in the same environment. This will usually involve some form of genetic distinction that discourages mating between two groups within a species.

EVOLUTIONARY CONSTRAINTS

the 30-second evolution

3-SECOND THRASH
Natural selection is powerful, but not all-powerful; not everything can evolve, but the limits on evolution differ between species.

3-MINUTE THOUGHT
Evolutionary biologists use the metaphor of an adaptive or 'fitness' landscape: a vista of hills and valleys where hills represent well adapted combinations of traits and valleys poorly adapted ones. Natural selection can only move a population uphill, never down. Genetic and developmental pathways can make a particular destination inaccessible, either because the path crosses a deep valley and intermediate forms would die or because the mutations needed cannot occur and there is no path at all.

The laws of physics explain why there are no flying pigs – or at least why an animal the weight of an adult pig could never fly. However, it's less simple to explain why there are no winged spiders or branching palm trees. A constraint is a limit on the outcomes of evolution due to physics or biology – or both. Flatworms are flat because oxygen cannot diffuse far through tissues (a physical reason) and they cannot take up oxygen any other way (a biological reason). By evolving lungs or other breathing organs, other animals freed themselves from this constraint. Other evolutionary constraints are more mysterious. Birds have varying numbers of vertebrae in their necks, with swans having up to 25, but in almost all mammals, from gerbils to giraffes, the number is fixed at seven. The reason for this is unclear, but probably lies in embryology: mutant mammals with extra neck vertebrae are common, but have additional deformities and rarely live to reproduce. Changes in the environment can also affect constraints. For example, when the ancestors of whales took to the water, they were liberated from the constraints of gravity, and were free to grow. They grew larger than any land animal ever could.

RELATED TOPICS
See also
GENETIC VARIATION
page 64

TYPES OF SELECTION
page 68

3-SECOND BIOGRAPHY
SEWALL WRIGHT
1889–1988
American geneticist who conceptualized the adaptive landscape

30-SECOND TEXT
Louise Johnson

The 'adaptive landscape' of evolutionary biologists maps how natural selection leads to the development of some traits and rejection of others.

COEVOLUTION

the 30-second evolution

RELATED TOPICS
See also
THE NEED FOR ADAPTATION
page 58

THE PARADOX OF SEX
page 118

SEX & EVOLUTIONARY
ARMS RACES
page 130

3-SECOND BIOGRAPHIES
CHARLES DARWIN
1809–82
English naturalist who first
mentioned the coevolution
concept in *On the Origin of
Species*

PAUL R. EHRLICH & PETER
H. RAVEN
1932– & 1936–
American co-authors of
groundbreaking work on
coevolution in relation to
plants and butterflies and one
of the first to use the term

3-SECOND THRASH
Evolution is sometimes
driven less by competition
within a species and
more by a linkage of
evolutionary change
between hosts and
parasites, predators and
prey and mutually
supportive organisms.

3-MINUTE THOUGHT
Some suggest we can go
beyond living systems for
coevolution, arguing that
the Earth's geology and
life could be deemed one
huge feedback loop. Some
2.3 billion years ago
photosynthetic bacteria
released vast amounts of
oxygen, changing the
chemical structure of the
air above us and rocks
beneath, in turn altering
the evolutionary trajectory
of life. Organisms still drive
enormous physical and
chemical changes in the
world, with people the latest
planetary bioengineers.

Evolution is often a response to
the environment in which the organism lives, yet
that environment includes other organisms, and
in coevolution an evolutionary change in one
organism is caused by a change that occurs in
something else that shares that environment.
Often this is at the level of mutually influencing
organisms, though coevolution can refer to
changes in two parts of an organism influencing
each other. The clearest examples of coevolution
tend to be between predator and prey, host and
parasite or organisms that have a symbiotic
relationship, though any organisms sharing an
environment and competing for resources can
experience coevolutionary forces. Orchids that
have developed elongated flowers as the moths
that traditionally pollinate them evolved ever-
longer proboscises demonstrate a coevolutionary
process driven by mutual dependence. When
coevolution drives an organism to develop a
new resistance to predation or parasitic attack
it can result in an unusually fast production of
new species, a process known as 'escape and
radiation coevolution'. Coevolution is a concept
often used beyond biology. Businesses
competing for the same customers must evolve
new products and services or ways of working,
while coevolution clearly drives the development
of computer hardware and software.

30-SECOND TEXT
Brian Clegg

*Orchids and the
moths that pollinate
them have coevolved
for more efficient
pollination.*

CONVERGENT EVOLUTION

the 30-second evolution

3-SECOND THRASH
All other things being equal, we expect closely related species to be similar. Convergent evolution occurs when all other things are not equal. Sometimes distantly related species appear alike as natural selection affects them in the same way, driving them to find similar solutions to the same problems.

3-MINUTE THOUGHT
While some instances of convergent evolution are obvious, most are difficult to detect as we need to see into the past to determine the nature of the ancestor of a species or group of species. The fossil record can at times facilitate this, but very often scientists use phylogenetic trees in conjunction with characteristics of contemporary species to infer what the past was like.

At first glance one could be forgiven for thinking that a dolphin is a fish. Of course, dolphins are actually mammals that are only distantly related to fish. Both have independently evolved a highly streamlined form, fins and tails that make them efficient and effective at locomotion in water. This physical adaptation to the aquatic environment is an example of convergent evolution. Convergence is characterized by species overtly sharing very similar characteristics, when those characteristics are not similar by descent. The ancestors of dolphins were land-dwelling deer-like animals that were well adapted for a terrestrial lifestyle and lived around 60 million years ago. In contrast, the ancestors of fish lived in the sea well over 500 million years ago. Convergent evolution has been common as distantly related species across the globe, and through geological time, have evolved to find solutions to similar problems. The wings of bats, birds and pterosaurs demonstrate this. These successful groups all evolved flapping flight from vertebrate terrestrial ancestors in very different ways. Convergence does not only occur at a morphological level. Recent scientific studies have demonstrated widespread convergent evolution at the level of genes between bats and dolphins which are distantly related echolocating mammal species.

RELATED TOPICS
See also
TRANSMUTATION AND ARCHETYPES
page 18

BUILDING PHYLOGENIES
page 38

FROM ADAPTATION TO SPECIATION
page 48

SPECIES DIVERSITY
page 50

NEW SPECIES
page 106

30-SECOND TEXT
Chris Venditti

Species that look similar are not always close relatives – physical adaptation to a shared environment has given them similar features despite a very different inheritance.

INDUSTRIAL MELANISM

the 30-second evolution

We are used to evolution being a slow process, but in short-lived creatures there can be quicker results, which enabled the peppered moth (*Biston bitularia*) to become a poster Lepidoptera for evolutionary theory. 'Melanism' refers to the pigment melanin, which occurs in the skin as a protection against the cell damage caused by ultraviolet light. A sun-tan, for example, is an increase in melanin level. The peppered moth's colouring enabled it to blend well with lichen-covered trees. However, during the Industrial Revolution, pollutants killed the lichen, exposing darker bark in trees that were also blackened directly by soot. Dark moths in the population were better camouflaged when resting on a blackened tree, so had a better chance of surviving the attacks of predators and hence reproducing. As a result, natural selection over generations produced more dark moths. In areas in which clean air laws have since reduced air pollution, the trend has reversed, giving the dark peppered moths less survival capability than the naturally lighter moths; the result is a return to the traditional lichen-like coloration of peppered moths. The same effect is visible in other moths and a handful of other insects, including at least one species of ladybird.

RELATED TOPICS
See also
VARIATION & SELECTION
page 20

FROM ADAPTATION TO
SPECIATION
page 48

THE NEED FOR ADAPTATION
page 58

3-SECOND BIOGRAPHIES
BERNARD KETTLEWELL
1907–79
British lepidopterist whose work comfirmed that lighter moths in polluted areas were more likely to be caught

J.B.S. HALDANE
1892–1964
British biologist who used a simple mathematical model to show that the changes in the moths occurred too quickly to be caused by random processes alone

30-SECOND TEXT
Brian Clegg

White or black pepper? The peppered moth exhibited very swift evolutionary changes in responding to pollution effects.

3-SECOND THRASH
Some insects have undergone colour changes that demonstrate the impact of natural selection when pollution changed the environment in which the insects live.

3-MINUTE THOUGHT
Despite very good evidence of correlation between changes in pigmentation and levels of pollution, it has not been definitively proved that the change in coloration of tree bark is the direct cause. There may be other reasons why an increase in pollution could result in a change in pigment levels – if, for instance, the melanin helped defend the moths against a toxin. However, there seems little doubt that this is an evolutionary response by means of natural selection to industrial pollution.

NEW SPECIES

the 30-second evolution

3-SECOND THRASH
When new environmental niches arise it is possible for species to form in short timescales, as demonstrated by the 500+ species of cichlid which have evolved in Lake Victoria in Africa.

3-MINUTE THOUGHT
The apple maggot fly (*Rhagoletis pomonella*) provides another example of rapid speciation in progress. It originally lived on the American hawthorn, but in the 1860s began to attack introduced apple trees and within decades evolved differing behaviours, driven by genetic differences that predispose its larvae to emerge when apples ripen. As yet both hawthorn and apple variants remain within *Rhagoletis pomonella*, but the divergence to separate species is underway.

New species often form due to geographic separation (allopatric speciation), but rapid speciation can occur for members of a single species that share a geographic area (sympatric speciation). Such 'adaptive radiation' is triggered when a species either encounters a new environment with a wide range of different environmental opportunities, or where a species evolves a new feature that enables it to use a wider range of existing environmental niches. The effect can be reinforced if appearance changes result in sexual selection, so newly forming species that could still interbreed do not. Rapid development of new species is common after a mass extinction, but for the cichlid fish of Lake Victoria, the formation of a large, isolated lake provided new opportunities, resulting in more than 500 species arising from a single parent species in a short evolutionary timespan. This was originally thought to be around 12,400 years ago, reflecting the date when Lake Victoria last dried out, but DNA evidence suggests that speciation began at least 100,000 years ago, kick-starting the process that flourished when the lake originally formed.

RELATED TOPICS
See also
MECHANISMS OF ISOLATION
page 44

FROM ADAPTATION TO
SPECIATION
page 48

SPECIES DIVERSITY
page 50

SEXUAL SELECTION
page 122

3-SECOND BIOGRAPHIES
SVEN OSCAR KULLANDER
1952–
Swedish biologist whose primary speciality is the study of cichlids and their speciation

30-SECOND TEXT
Brian Clegg

Highly evolved ... the cichlid fish and the apple maggot fly underwent significant changes in a short timespan.

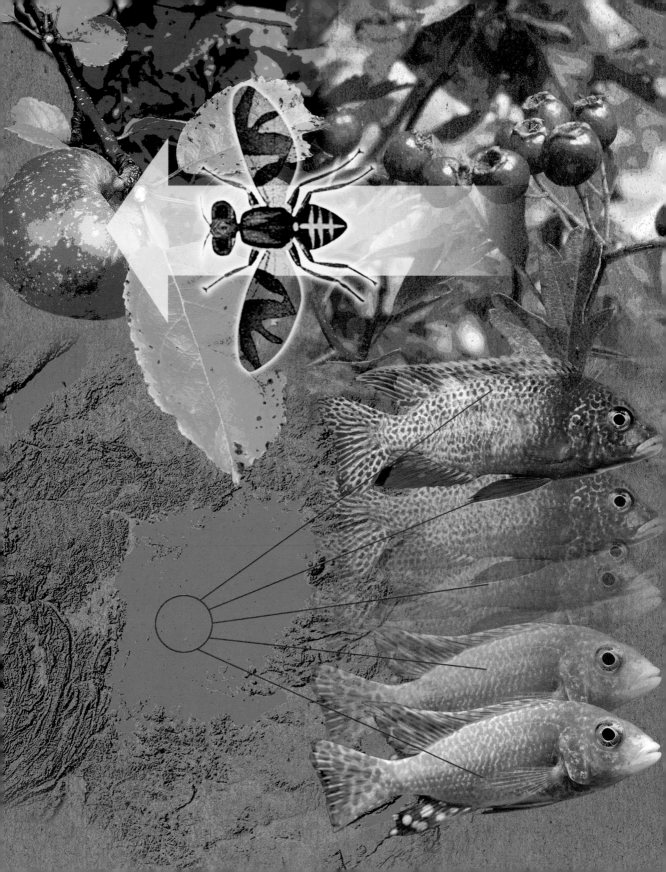

8 October 1936
Barbara Rosemary Matchett born in Arnside, Cumbria

26 October 1936
Peter Raymond Grant born in Norwood, London

1960
Peter receives BA from Cambridge University; Rosemary receives BSc from Edinburgh University

1964
Peter receives PhD from the University of British Columbia, Vancouver

1973
Peter becomes Professor at McGill University, Montreal

1973
The Grants make their first trip to the Galapagos island of Daphne Major

1977
Peter becomes Professor at University of Michigan

1985
The Grants move to Princeton University

1985
Rosemary receives PhD from Uppsala University, Sweden

1994
Jonathan Weiner's Pulitzer Prize-winning book about the Grants' work, *The Beak of the Finch*, published

2002
Jointly awarded Darwin Medal by the Royal Society

2009
Jointly receive the Darwin-Wallace medal from the Linnean Society of London

PETER & ROSEMARY GRANT

At the age of four, Peter Grant was evacuated from London to southern England, where he revelled in collecting butterflies and bird watching. Meanwhile, in rural Cumbria, Rosemary Matchett's interest in nature was encouraged by her mother who took her on field trips to discover fossilized plants.

After studying zoology and botany at Cambridge, Peter travelled to British Columbia, where he met Rosemary within days of starting his PhD in zoology. She had studied genetics at Edinburgh and delayed embarking on a PhD when the opportunity arose to lecture in Vancouver. They married about a year later. The Grants' focus became the interplay of ecology and evolution, exploring how the environment shapes distributions and properties of species. This led to work on bird beak sizes on the Tres Marías islands off Mexico, where they discovered that average beak sizes were larger than those on the mainland, giving the birds an advantage with the available foodstuffs. The discovery supported an idea the couple were developing; that inter-species competition for food influences evolutionary development.

Inspired by a book on Darwin's finches, the Grants travelled to Daphne Major in the Galapagos to explore how the 14 species of the bird found there arose. The island proved an ideal habitat to study the evolutionary pressures of competition, food supplies and environment. The couple have returned every year since their first visit in 1973. During a severe drought on Daphne Major in 1977, the Grants found that big-beaked birds, better able to crack open the large, hard seeds that were more plentiful after the drought, had a higher survival rate, producing more offspring with big beaks the following year: a clear case of evolution by natural selection. A few years later the position reversed when heavy rainfall boosted small-seeded plants, making it easier for smaller-beaked finches to survive. They were also able to study how birdsong is influenced by competition and to observe the impact of the arrival of a new, larger-beaked species that became the dominant species living on larger seeds and forcing out most existing large-beaked birds.

Moving to Princeton in the 1980s, the Grants adopted a more statistical approach, predicting the degree of evolutionary change likely from a particular parental generation, and collaborating on work on the genetic factors involved in beak shapes and sizes. The value to biologists of being able to predict evolutionary change cannot be overstated, a point not lost on geneticist J.B.S. Haldane, who once remarked: 'No scientific theory is worth anything unless it enables us to predict something which is actually going on. Until that is done, theories are a mere game of words,' More recently, Rosemary has focused on hybridization between finch species and its potential for producing evolutionary advantages and disadvantages.

Brian Clegg

EVOLUTION OF ANIMAL BEHAVIOUR

the 30-second evolution

3-SECOND THRASH
Studies of animal
behaviour confirm that
behaviours – just as much
as physical characteristics
– can evolve and be
adapted to different
environmental pressures.

3-MINUTE THOUGHT
Social behaviour can seem
counter-intuitive from an
evolutionary standpoint,
but in reality social
behaviours increase the
ability to survive and pass
on genes. Take the group
behaviour of prey animals,
for example. Solitary
wildebeests would tend
to be picked off by
predators like lions, but
by forming a herd they
reduce the chance of an
individual being attacked.
Despite some costs –
reduced grazing, for
instance – adaptation to
life in a group brings
survival benefits.

Ethology, the study of animal
behaviour, has evolution at its heart and Darwin
made an early venture into this field, though it
was another 60 years before it really took off.
Konrad Lorenz proposed 'fixed action patterns',
instinctive behaviours controlled by an 'innate
releasing mechanism' in the brain, triggered by
an external influence, with examples ranging
from mating dances to the automatic response
of birds to return lost eggs to the nest. Nikolaas
Tinbergen emphasized the importance of
instinctive behavioural reactions that are driven
by evolutionary and adaptive mechanisms. Often
the environment is key to such reactions; for
instance, the ungainly, bulging vocal sacs of
many frogs, which resonate to amplify mating
calls, tend not to be found in environments with
loud background sounds, like noisy running
water. Differing parental care – from laying large
numbers of eggs without support to some
mammals' extended parental care – is influenced
by ecology as much as biological factors. From
the 1970s behavioural studies took a wider view,
more concerned with the social aspects of
animal behaviour and so giving emphasis to the
evolution of group behaviour, from the complex
'eusocial' superorganisms of ants and bees to
smaller mammalian social groupings.

RELATED TOPICS
See also
THE NEED FOR ADAPTATION
page 58

ALTRUISM & SELFISHNESS
page 112

EVOLUTIONARY PSYCHOLOGY
page 146

3-SECOND BIOGRAPHIES
KARL VON FRISCH
1886–1982
Austrian biologist and early
ethologist, best known for
work on bees, who shared the
1973 Nobel Prize for
discoveries in behaviour

KONRAD LORENZ
1903–89
Austrian zoologist and early
ethologist, best known for
work on imprinting who shared
the 1973 Nobel Prize for
discoveries in behaviour

30-SECOND TEXT
Brian Clegg

*From mating dances to
animals or fish finding
safety in numbers,
instinctive behaviour
is a response to the
environment.*

ALTRUISM & SELFISHNESS

the 30-second evolution

3-SECOND THRASH
Considering purely the individual, altruism seems counter-intuitive, if the sole urge is selfish preservation of an individual's genes – but evolutionary theory explains the potential benefits of altruism.

3-MINUTE THOUGHT
Altruism in humans is widespread, yet there's no suggestion that we coldly calculate our relatedness to the recipient, or our likelihood of receiving a benefit for acts of generosity. Nevertheless, we generally treat family better than non-family, friends better than outsiders, countrymen better than foreigners. Explaining the kindness of strangers is a significant challenge for evolutionary psychologists.

Viewed simplistically, altruism – putting others first – doesn't make evolutionary sense. Yet there are several ways that altruism can develop within an evolutionary framework. It has frequently been explained as a result of kin selection, the idea being that close relations share some genetic heritage, so apparent altruism is still selfish when considering the preservation of the gene pool. However, it seems impossibly calculating – as J.B.S. Haldane jokingly said 'I'd lay down my life for two brothers or eight cousins' – beyond a simple tendency to protect close family, genetically related or not. The picture becomes more convincing when reciprocal altruism is added, where an individual is prepared to undertake acts of no personal benefit on the understanding that they can expect similar altruism from others, effectively making selfishness mutual. (This comes through strongly in the biblical petition 'forgive us our trespasses as we forgive those who trespass against us'.) Just as paper money replaced silver and gold, in our complex modern world reputation has replaced direct experience as a means of judging others' trustworthiness. This allows indirect reciprocity to develop even if no altruistic act has been performed to benefit another individual.

RELATED TOPICS
See also
MECHANISMS OF ISOLATION
page 44

FROM ADAPTATION TO SPECIATION
page 48

SPECIES DIVERSITY
page 50

BILL HAMILTON
page 128

3-SECOND BIOGRAPHIES
EDWARD OSBORNE WILSON
1929–
American biologist who promotes the concept of group selection – but few evolutionary biologists agree

ROBERT L. TRIVERS
1943–
American evolutionary biologist who developed the concept of reciprocal altruism

30-SECOND TEXT
Brian Clegg

Behave considerately on the understanding that others will do the same.

SEX & DEATH

SEX & DEATH
GLOSSARY

Asexual reproduction Reproduction in which an organism only has a single parent and thus is a genetic copy or clone of that parent, carrying only genes from that parent. Includes parthenogenesis, fission (splitting into two), sporulation and fragmentation.

Beneficial mutation A change in a genome, typically produced by damage to DNA or a replication error, producing a difference in the organism's characteristics that increases its chance of survival and thus is likely to be favoured by natural selection.

Counter adaptation An adaptive response from a natural enemy to a defensive adaptation by its prey (and vice versa).

Cuckoldry Taken from Old French, cuckold referred to the husband of an unfaithful wife, a name based on the cuckoo and its habit of laying its eggs in the nest of another species, or another member of its own species. Cuckolded males expend energy rearing the offspring of another male, often also forgoing the chance to rear their own offspring.

Eusociality A type of animal behaviour involving different subgroups with distinct roles, and a shared brood, usually produced by a single 'queen'. The subgroups often lose the abilities required for other roles. Mostly ants, bees, wasps and termites, though mole rats are eusocial mammals. 'Eusocial' means 'good social' implying that it is the most organized of the social forms.

Evolutionarily stable strategy A concept originating in game theory, a population that has an evolutionarily stable strategy is one in which natural selection ensures that the population cannot be displaced by a mutation that is initially scarce.

Fit/fittest/fitness Being well adapted or suited to conditions. In the evolutionary sense 'survival of the fittest' refers to those best suited to survive and pass on genetic material.

Inbreeding/inbreeding depression When individuals that are closely genetically related mate the result is described as inbreeding, which produces an increased chance of passing on traits that will decrease the biological fitness of the offspring. When applied to a population the overall decline in fitness is described as inbreeding depression.

Kin selection The idea that evolutionary processes can work to the benefit of related organisms even if they are to the detriment of an individual, hence explaining altruistic behaviour in which an individual makes a sacrifice to benefit its kin. Most clear in eusocial organisms that lose the ability to reproduce.

Major Histocompatibility Complex (MHC) A set of molecules protruding from the surface of a cell that T-cells, a form of white blood cell, lock onto to 'read' the internal make-up of the cell and decide if it should be ignored or destroyed.

Mutational meltdown A spiral of increasing mutations within a population, in which harmful genetic changes produce a decline in the population, which makes further negative mutations more likely to accumulate.

Parthenogenetic reproduction Literally 'virgin birth', a form of reproduction in which an embryo grows from an unfertilized egg.

'Red Queen' hypothesis The idea that – like Lewis Carroll's character the Red Queen in *Through the Looking-Glass* who had to keep running simply to stay in the same place – organisms need to evolve in order to survive when the environment, competitors and predators are themselves evolving.

Relatedness The degree to which an individual is related to another (how much of a genome they share), which influences the degree to which parent-offspring conflict arises, so that, for instance, two half brothers would be expected to generate a stronger conflict than two full brothers.

Self-incompatability A genetic mechanism preventing a plant from fertilizing itself, encouraging genetic diversity.

Sex-biased dispersal As part of their life cycle some organisms disperse from their birthplace to gain potential evolutionary benefits. Many species practise sex-biased dispersal, where one sex tends to breed near its birth site while the other disperses to a new breeding site.

THE PARADOX OF SEX

the 30-second evolution

3-SECOND THRASH
Sex maintains genetic variation, avoiding the accumulation of deleterious mutations and allowing the accumulation of beneficial ones.

3-MINUTE THOUGHT
Sex appears costly, but introduces genetic novelty, the grist to the mill of natural selection. In many ways it is not the existence of sex that needs explaining, but rather how some asexually reproducing groups appear to have persisted for huge amounts of evolutionary time without apparent change. Aquatic microorganisms called bdelloid rotifers appear to have reproduced parthenogenetically for at least 35 million years; do they have means of repairing their genomes that sexually reproducing species lack?

Just what is the point of males?

Many species get along perfectly well without them. They reproduce asexually, producing clones of themselves. Unless something stops them, asexual species have populations that can grow exponentially, like a nuclear chain reaction. One, two, four, eight, sixteen ... After another 15 generations, only a few days for some species, they have more than 1 million near-identical descendants. Sexually reproducing species appear to squander energy on useless males, who are unable to reproduce. If sexual reproduction is so wasteful, why bother with it? There are a number of reasons, but all argue that sex reduces the risk of extinction. One idea is that in small populations the recombination of genes during sexual reproduction means that deleterious mutations are flushed out of the genome, whereas asexual species accumulate them. This gain in deleterious mutations is irreversible and leads to 'mutational meltdown' and extinction. Another key idea is that the continual mixing of new combinations of genes makes this cost of males worthwhile. If a beneficial mutation arises in an asexual individual, it is stuck there, only passed to its offspring. Sexual reproduction allows the mingling of different beneficial mutations in one genome, enabling rapid adaptation to evolutionary pressures.

RELATED TOPICS
See also
SEX RATIOS
page 120

SEXUAL SELECTION
page 122

SEX & EVOLUTIONARY ARMS RACES
page 130

INBREEDING AVOIDANCE
page 132

3-SECOND BIOGRAPHY
HERMANN JOSEPH MULLER
1890–1967
American Nobel laureate, renowned for his work on the effect of radiation on rates of genetic mutation

30-SECOND TEXT
Mark Fellowes

Male and female ... sexual, rather than asexual, reproduction allows for swifter response to evolutionary demands because it boosts genetic variation.

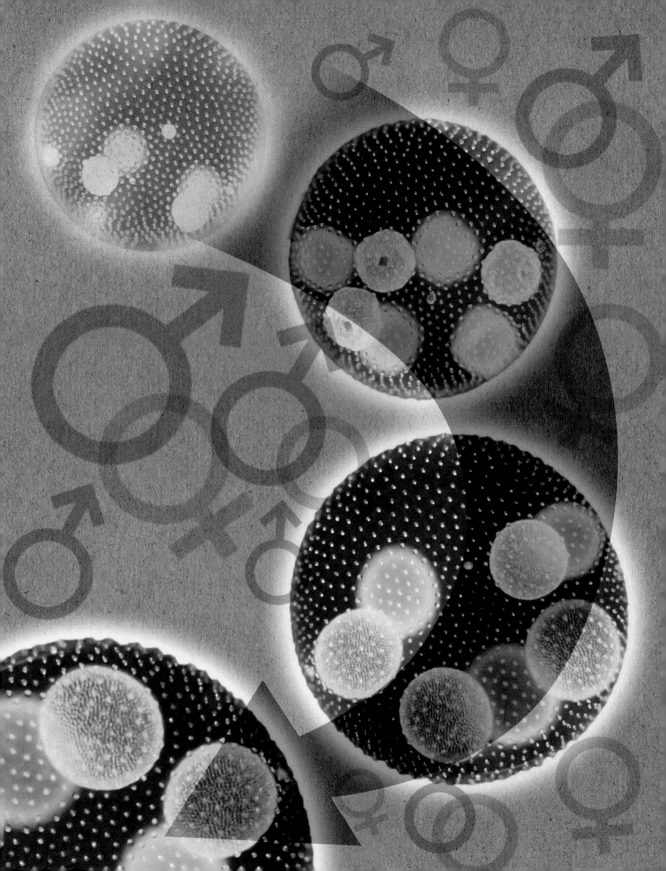

SEX RATIOS

the 30-second evolution

While males are necessary for sexual reproduction, why do we see so many of them? A single male can fertilize many females, so a species producing equal numbers of males and females seems to be wasting resources and effort. Working in the early 1930s, Ronald Fisher realized that while every sexually reproducing species has to have a mother and a father, there is an advantage to being the rarer sex. Imagine a population in which there are two females for each male. Each male on average will mate with two females, having double the fitness of a typical female. The genes for producing males will spread and the benefit of being male is eroded as a 1:1 sex ratio is approached. The same works if females are rare, except that now some males will fail to gain matings, so on average being female will result in more offspring. This is what is known as an evolutionary stable strategy, a point at which natural selection prevents other strategies from taking over. In humans, the natural birth sex ratio is slightly male-biased (around 106 males for every 100 females born), but as males generally die earlier than females this difference is reduced over time.

RELATED TOPICS
See also
THE PARADOX OF SEX
page 118

BILL HAMILTON
page 128

INBREEDING AVOIDANCE
page 132

3-SECOND BIOGRAPHY
RONALD FISHER
1890–1962
English statistician Richard Dawkins called the 'greatest evolutionary biologist since Darwin' for work linking genes and natural selection

30-SECOND TEXT
Mark Fellowes

3-SECOND THRASH
Even patterns as ubiquitous in nature as the 1:1 sex ratio of most species are explained by natural selection.

3-MINUTE THOUGHT
Not all species have a 1:1 sex ratio. Fig wasps can choose their offspring sex; fertilized eggs become female, unfertilized male. It is not uncommon to find one male for every 19 female fig wasps in a fig fruit (syconium). As a single female lays all the eggs in a syconium, the males and females will all be related. Since the males are genetically similar, the parent female will have more grand-offspring if she chooses to produce just enough males to fertilize all of their sisters.

Inside the fig the male fig wasp's task is to fertilize the females. He may be outnumbered 19 to 1.

SEXUAL SELECTION

the 30-second evolution

Darwin realized that selection was not just about who lives, but also about who breeds. He introduced sexual selection as a complementary idea to natural selection, to describe how variation in reproductive success could lead to some of the most extravagant features and behaviours of the living world. Inherent in concepts of sexual selection is the idea that one sex, usually the female, invests more in reproduction, so she should work to select the best males to mate with. In contrast, males can mate with many females, and therefore the males compete with each other through features that increase their chances of winning confrontations. Here we see the costly weaponry of the red deer or the enormous bulk of male elephant seals deployed in savage battles, physical traits which help males win confrontations. With female choice the characteristic trait preferred may have no apparent relationship to fitness at all, and can lead to runaway selection for 'ornaments' such as the incredible tail of the male peacock. By selecting such males, she in turn will have 'sexy sons', who are more likely to pass on her genes. Others have suggested that such displays are not accidents, but instead provide honest advertisements of health, allowing females to choose the fittest males.

3-SECOND THRASH
The concept of sexual selection explains many of the most complex behaviours and flamboyant traits found in nature.

3-MINUTE THOUGHT
Has sexual selection influenced the evolution of human physical and behavioural traits? It would be surprising if this were not so and some have suggested that the human brain is the result of sexual selection, with intelligence a favoured characteristic. There are more controversial studies, which suggest that female mate preference varies during her ovulation cycle, with dominant males preferred when she is most fertile.

RELATED TOPICS
See also
THE PARADOX OF SEX
page 118

SPERM COMPETITION
page 124

SEX & EVOLUTIONARY ARMS RACES
page 130

3-SECOND BIOGRAPHIES
AMTOZ ZAHAVI
1928–
Israeli evolutionary biologist who introduced the concept of 'honest advertisements', helping to explain why sexual selection could result in exaggerated features

MARLENE ZUK
1956–
American behavioural ecologist who helped develop the idea of extravagant ornaments being linked to male quality

30-SECOND TEXT
Mark Fellowes

Sexual selection lies behind displays of a peacock's finery and a red deer or elephant seal's aggression.

SPERM COMPETITION

the 30-second evolution

Why do males produce so many sperm? The traditional view was that sperm is cheap, so there is no evolutionary pressure to reduce the amount produced. This perspective was challenged by Geoff Parker in the early 1970s, who noted that sperm quantity, like other biological traits, was optimized through natural selection. Even among our closest relatives, testes size (a proxy for the volume of sperm produced) varies greatly. Gorillas have relatively small testes, orangutans slightly larger, then humans and finally chimpanzees. If a female mates with two males, the male with the most sperm is likely to have the greatest chance of siring offspring. The strength of selection for sperm quantity is related to mating behaviour. With gorillas, a single silverback male dominates matings in his group, making sperm quantity less critical. Chimpanzees are relatively promiscuous, with several males mating with a female. This leads to sperm competition, explaining their relatively large testes. Human testes are intermediate in size, suggesting that we lie somewhere in the middle of the sperm competition continuum. Recent genetic studies suggest that around 1 per cent of children are the result of cuckoldry, consistent with the idea that sperm competition is an unusual, but not exceptional, part of human reproductive behaviour.

3-SECOND THRASH
Competition does not end with copulation, it continues until the egg is fertilized.

3-MINUTE THOUGHT
Humans are socially monogamous: males and females form long-term pair bonds but may also pursue extra-pair matings. The cost of extra-pair copulation is potential cuckoldry, where the male partner bears the cost of rearing another male's offspring and losing the opportunity to reproduce himself. Natural selection has worked to combat this cost: the shape of the human penis is thought to help remove sperm from previous matings, and males whose female partners may have copulated with another male produce more sperm and attempt to copulate more frequently.

RELATED TOPICS
See also
SEXUAL SELECTION
page 122

SEX & EVOLUTIONARY ARMS RACES
page 130

INBREEDING AVOIDANCE
page 132

3-SECOND BIOGRAPHY
GEOFFREY ALAN PARKER
1944–
British behavioural ecologist who first suggested the concept of sperm competition, identified through his work with dung flies

30-SECOND TEXT
Mark Fellowes

A single male silverback gorilla has control over his group's mating so unlike chimpanzees (and men) does not need to compete in terms of quantity of sperm.

PARENT-OFFSPRING CONFLICT

the 30-second evolution

3-SECOND THRASH
Parents and their offspring have differing opinions over the optimal allocation of parental resources.

3-MINUTE THOUGHT
Geneticist David Haig suggested that there is evidence of parent-offspring conflict in humans, with foetuses demanding more resources from their mothers than their mothers wish to give. This conflict is played out through hormones secreted by the developing baby's placenta, which act to raise blood sugar levels; this is answered by the mother producing more insulin to reduce them. Perhaps as a mother ages and so is less likely to have future offspring, her strategy should change to approach that of her offspring.

In sexually reproducing species there is an inevitable evolutionary battle between parents and offspring. Robert Trivers recognized that parents should balance their investment in individual offspring to maximize the overall number and quality of their progeny. The offspring should not want this investment shared because it might reduce their own chances of reproducing successfully. This can be explained by relatedness. Parents, being equally related to all their offspring, should invest equally. While offspring carry 100 per cent of their own genes they share 50 per cent with their siblings so – from a gene-centric standpoint – they should attempt to obtain more resources from their parents than the parents should be willing to give, increasing their survival chances at the cost of their current or future siblings' survival. This explains why many offspring actively resist parental attempts to wean them. Chimpanzee mothers often have to forcibly discourage their offspring from breastfeeding when they become fertile again and begin to seek new matings. Competition between offspring for parental resources can become so extreme that it ends in siblicide. This is most noticeable with birds of prey, where in many species the elder sibling will kill the weaker, younger sibling, particularly when food is scarce.

RELATED TOPIC
See also
ALTRUISM & SELFISHNESS
page 112

3-SECOND BIOGRAPHIES
ROBERT TRIVERS
1943–
Influential American biologist who revolutionized our understanding of cooperation and conflict

DAVID ADDISON HAIG
1958–
Australian geneticist who first suggested maternal-foetal conflict and provided the basis for a new understanding of some of the most threatening complications of pregnancy

30-SECOND TEXT
Mark Fellowes

Needy young ... in humans mother and foetus hormones play out a battle over blood sugar, while in the nest often only the fittest survive.

1 August 1936
Born in Cairo to New Zealand-born civil engineer Archibald Hamilton and doctor Bettina Hamilton

1964
Lectureship at Imperial College, London and publication of key paper on the genetic evolution of social behaviour

1966
Marries Christine Friess, with whom he had three daughters

1970
Publishes key paper on Hamiltonian spite

1976
Publication of Richard Dawkins's *The Selfish Gene*, popularizing Hamilton's theory

1978
Becomes professor of evolutionary biology at the University of Michigan

1980
Elected Fellow of the Royal Society

1984
Becomes professor at Oxford University

1988
Awarded Darwin Medal of the Royal Society

1989
Awarded Scientific Medal of the Linnean Society

1993
Awarded the Crafoord Prize (the biology equivalent of a Nobel)

1994
Meets Luisa Bozzi, who becomes his partner

7 March 2000
Dies in London

BILL HAMILTON

Mention 'the selfish gene' and we think of Richard Dawkins, but the title of Dawkins's book paid tribute to William Hamilton. Though born in Egypt to New Zealander parents, Hamilton was brought up in Kent, where he developed an interest in butterflies, and apart from a brief spell in the US spent the rest of his life in the UK. When studying at Cambridge, Hamilton developed an interest in the statistically based genetics of Ronald Fisher, leading to a PhD for which he was jointly enrolled at the London School of Economics and University College, London. He spent thirteen years as a lecturer at Imperial College, where his research was considered far better than his lecturing.

Hamilton was still in his 20s when he published the two-part paper 'The Genetical Evolution of Social Behaviour', in which he posited 'Hamilton's rule', taking a quantitative approach to kin selection, linking the closeness of relationship to the cost of being altruistic. Such kin selection had been hypothesized already, but Hamilton gave it numerical rigour.

Other key concepts on which Hamilton worked were Hamiltonian spite and extraordinary sex ratios. Spite is, biologically, the antithesis to altruism, the idea that there is a genetic survival benefit to harming those who are less similar to yourself than the average population. Such 'spiteful' behaviour does exist – for example, when animals kill the infants of their rivals – but the concept has never had the same level of support as altruism in evolutionary psychology. Hamilton's work on extraordinary sex ratios looked at cases in which there is a wide divergence from 'ordinary' sex ratios of roughly 1: 1 – as in ants and wasps. Here once again he was introducing more mathematics to biology than was common, using game theory to explain the stability of extraordinary sex ratio communities. Hamilton spent much of his later career studying parasites, which he believed were important in the evolution of sex.

After brief visiting professorships at Harvard and São Paulo, Hamilton spent six years as professor of evolutionary biology at the University of Michigan before retuning to the UK to take up a research professorship in the department of zoology at Oxford, becoming a fellow of New College, where he remained until his death. This was widely attributed to malaria contracted during an expedition to the Congo, but was the result of gastrointestinal haemorrhaging, possibly from a pill becoming lodged in a pouch in the wall of the duodenum.

Brian Clegg

SEX & EVOLUTIONARY ARMS RACES

the 30-second evolution

Predators winnow out the weak
and the less well-adapted individuals, gradually improving the defence traits of their prey. As prey become quicker, stronger, more camouflaged, better defended, the poorer and less-effective hunters are lost, in turn selecting for superior predators. Each process causes a positive feedback loop of adaptation and counter-adaptation. Leigh van Valen's 'Red Queen' hypothesis connects these arms races with sex. The Red Queen in Lewis Carroll's *Alice in Wonderland* says that 'it takes all the running you can do, to keep in the same place'. Van Valen suggested that this was just like life, where the continual change allowed by sex is needed just to keep up with ever-changing enemies. Species must continually transform their immune response if they are to keep ahead of their rapidly reproducing diseases and parasites, which in turn have to break through the host's immune response if they are to survive. Where there is no variation in immune response, the right key will always open all locks. Sexual reproduction maintains variations in genes, continually shuffling the pins in the lock, so the diseases must also try different keys.

3-SECOND THRASH
Sex allows rapid adaptation in response to parasites and diseases, which would otherwise overwhelm us.

3-MINUTE THOUGHT
Can we see arms races between the sexes? If males maximize fitness by mating multiple times, but mating is costly to females, we could expect to see an arms race. Mating can be damaging for females of some species, with males using traumatic insemination (bedbugs), love darts (snails) and poisonous semen (fruit flies) to maximize their chances of success. In response females try to reduce the cost of mating, but few species have taken it as far as the praying mantis, where females practise sexual cannibalism.

RELATED TOPICS
See also
COEVOLUTION
page 100

THE PARADOX OF SEX
page 118

SEXUAL SELECTION
page 122

3-SECOND BIOGRAPHY
LEIGH VAN VALEN
1935–2010
American evolutionary theorist who argued that species evolved through constant arms races, continually adapting to their enemies and victims

30-SECOND TEXT
Mark Fellowes

Sex can be dangerous. The female praying mantis bites the head off the male during mating; snails fire a love dart into their mate as part of courtship.

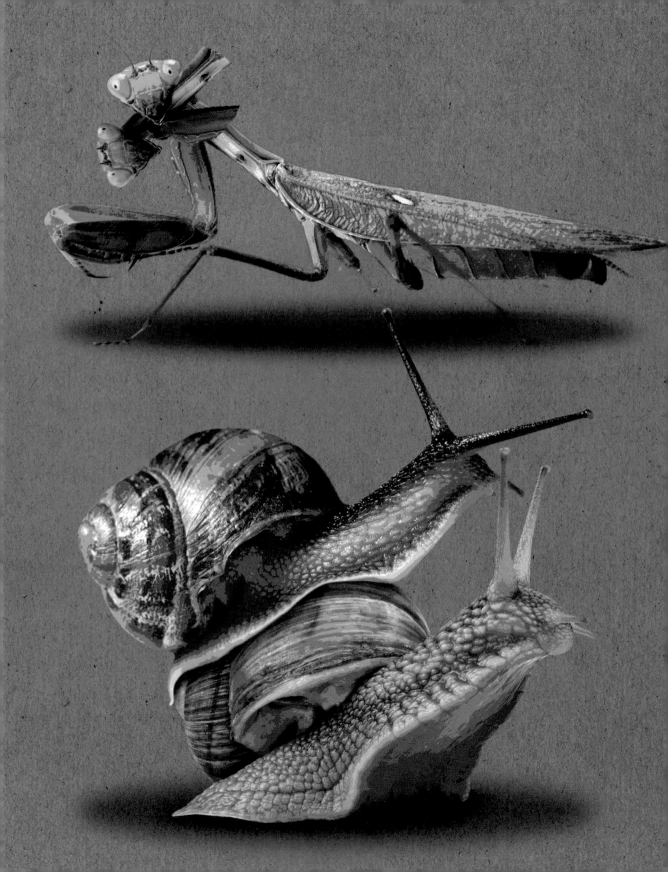

INBREEDING AVOIDANCE

the 30-second evolution

RELATED TOPICS
See also
THE PARADOX OF SEX
page 118

SEXUAL SELECTION
page 122

SPERM COMPETITION
page 124

30-SECOND TEXT
Mark Fellowes

3-SECOND THRASH
Inbreeding can be highly costly and many species have evolved traits that reduce the likelihood of mating with relatives.

3-MINUTE THOUGHT
Controversial studies suggest that people prefer mates who smell different to themselves. This difference in odour is caused by the Major Histocompatibility Complex (MHC), which controls the expression of cell surface receptors and plays a key role in immunity. Intriguingly, it has been suggested that women taking the contraceptive pill prefer males with MHC similar to their own, while those not taking it prefer different odours.

Some sexually reproducing species can fertilize their own eggs. Plants frequently reproduce in this way. But inbreeding is not without costs. Self-fertilization or mating with close relatives increases the chances that the resulting offspring will receive two copies of a deleterious gene. The traits affected by these genes are likely to be seen, rather than hidden by a different form of the gene. This leads to inbreeding depression, as the bad genes act to reduce the fitness of their carriers. This has been seen in the royal families of Europe, where politically motivated inbreeding was widespread and the result was disease and deformity. Most species have evolved a range of mechanisms to minimize the chances of inbreeding. Plants may be self-incompatible, so their own pollen cannot fertilize their own ovules. This incompatibility is less common in animals, which have behavioural traits that minimize the chances of inbreeding. Many animal species reduce the likelihood of inbreeding by exhibiting sex-biased dispersal, where one sex remains where they were reared and the other disperses. Other species use direct cues. The scent of house mouse urine varies with the major urinary proteins expressed by the carriers and the mice avoid inbreeding by mating with those whose urine smells different to their own.

Inbreeding in the Royal House of Hapsburg led to a deformed chin – as seen in King Charles II of Spain (rear) and Holy Roman Emperor Charles V.

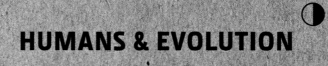

HUMANS & EVOLUTION

Artificial selection Selectively breeding organisms in order to enhance a particular trait or traits. Domestic animals and plants have undergone artificial selection to produce, for example, the huge range of variety seen in the single species of dog.

Australopithecines Early hominids from around 1.2 to 4 million years ago. A collection of hominin species that includes the genera *Australopithecus* and *Paranthropus*.

Bipedalism/bipedality Moving using only the two rear limbs. Most common amongst birds, which inherited the ability from dinosaurs, bipedalism is uncommon among mammals, but is the preferred method of locomotion for humans.

Genetic engineering/synthetic selection The use of technology to modify an organism's genes directly in order to produce new traits – as opposed to artificial selection, in which the modification is made indirectly.

Genetic modification (GM) An organism that is produced by genetic engineering is described as genetically modified. In the process genes can be added, deleted or modified. The term is usually not applied to products of selective breeding, even though technically these are also genetically modified.

Hominids The family of primates that in living species comprises the great apes: humans, chimpanzees, gorillas and orangutans. It also includes the extinct species that are closer to us than the chimpanzee.

Hominins Hominids that are genetically closer to human beings than chimpanzees – *Homo sapiens* is the only extant (living) species of hominin.

Homo erectus A hominin species that became extinct around 140,000 years ago. It is not clear if it is the same species as *Homo ergaster*; if it is, it is likely to be a direct ancestor of *Homo sapiens*.

Homo habilis A hominin species that flourished 2.3 to 1.4 million years ago. Least like modern humans of all the members of the genus *Homo*, but is usually placed in this genus because of its relatively large brain size.

Homo heidelbergensis A hominin species that lived from at least 600,000 years to perhaps 200,000 years ago. With a similar brain size to *Homo sapiens*, this species could be the direct ancestor of modern humans and Neanderthals.

Mitochondrial DNA analysis A small amount of DNA is found in the mitochondria, the so-called powerhouse in eukaryotic cells, providing just 37 genes in humans. In most species this DNA is inherited only from the mother, and comparison of mitochondrial DNA within populations can be used to deduce how those populations developed over time, while comparison between species helps clarify the way that the species have developed from common ancestors.

Neanderthals A species (*Homo neanderthalensis*) of hominin closely related to *Homo sapiens* that went extinct only 20,000 to 30,000 years ago. There is evidence of interbreeding with *Homo sapiens*, as modern humans carry a small percentage of Neanderthal DNA.

Nuclear DNA The DNA contained within the nucleus of a eukaryotic cell, comprising the vast majority of the DNA of most eukaryotes, but supplemented by mitochondrial and chloroplast DNA.

Palaeoanthropology The study of ancient hominins in fossil form.

Proconsul A genus of primates that went extinct around 14 million years ago. Once thought to be an ancestor of the great apes, though this is now considered more doubtful.

Spandrel Originally a term in architecture meaning the space in the corner of an arch, 'spandrel' was taken by evolutionary biology to mean an evolutionary change that happens as a side-effect of an adaptive change, but that may turn out to be useful in its own right.

ANCESTORS & TIMESCALES

the 30-second evolution

The oldest ancestors of modern humans (*Homo sapiens*) that are clearly distinguishable from other primates date back 6–7 million years and were found in Africa. These included two species, *Orrorin tugenensis* and *Sahelanthropus tchadensis*, which probably walked on two legs. But these Australopithecines still had apelike long arms and small brains. About 4 million years later, we find that brain size has increased and stone tool-making has begun. This change characterizes the birth of our own genus – *Homo*. Around 1.8 million years ago *Homo erectus* adopted a hunter-gatherer lifestyle and spread out of Africa and into Asia. The next few hundred millennia witnessed the evolution of the Neanderthals in Europe, the Denisovans in Asia and in Africa, about 200,000 years ago, *Homo sapiens*. This last species spread out of Africa and by 60,000 years ago reached Australia; 40,000 years ago they reached Europe and 15,000 years ago arrived in South America. Mitochondrial DNA analysis has confirmed that the mother of all modern humans (Mitochondrial Eve) came from Africa. Nuclear DNA also shows that, despite some interbreeding with local populations of Neanderthals and Denisovans, modern humans came out of Africa and gradually replaced all other humans that then inhabited Earth.

RELATED TOPICS
See also
TOOL USE BY HUMANS & OTHER APES
page 140

THE LEAKEY FAMILY
page 148

3-SECOND BIOGRAPHIES
LOUIS LEAKEY
1903–1972
British anthropologist who discovered *Homo habilis*; first human to make and use tools

CHRIS STRINGER
1947–
British anthropologist and one of the leading proponents of the Out of Africa theory

SVANTE PÄÄBO
1955–
Swedish evolutionary geneticist, known as one of the founders of paleogenetics

30-SECOND TEXT
Isabelle De Groote

Skulls of Neanderthals, Homo antecessors, Homo erectus and Homo sapiens. Africa is the original home of the human species.

3-SECOND THRASH
Our own species, *Homo sapiens*, evolved in Africa around 200,000 years ago and spread across the world replacing other humans that lived there.

3-MINUTE THOUGHT
When modern humans migrated out of Africa around 60,000 years ago, they found Europe occupied by Neanderthals. Because 1–4 per cent of Neanderthal DNA is present in modern Europeans we know that some encounters between the two species produced offspring. Recent analyses have shown that this interbreeding had a positive effect on immunity to European-specific diseases, but may also have introduced illnesses such as lupus, Crohn's disease and biliary cirrhosis.

TOOL USE BY HUMANS & OTHER APES

the 30-second evolution

What distinguishes us from other hominids (humans and their ape relatives) is the way humans make and use tools. Since the discovery of *Homo habilis* (or 'the handy man') by Louis Leakey in 1964 tool manufacture has been considered the hallmark of our own genus, setting it apart from earlier human species and the other apes. Increasingly, scientists are uncovering how important tool use is for humans and other hominids. Chimpanzees are the most frequent tool-users; they use spear-like weapons for hunting and stones for cracking nuts, and make different dipping sticks to collect ants, termites or honey. Orangutans use sticks to open prickly fruits and have been seen to use leaves to protect their hands from the thorns. Gorillas use sticks to test the depth of water when crossing swampy areas or as a walking stick. These observations reinforce the notion that the use of tools began long before humans evolved and was probably shared by the last common ancestor more than 12 million years ago. Nevertheless, humans are the only species that do such innovative things with tools. We constantly improve on their design, not just functionally – but aesthetically, too.

3-SECOND THRASH
Tool-making was considered a defining trait of humans, but the more we learn about apes the more blurred these definitions become.

3-MINUTE THOUGHT
Increasingly sophisticated tools meant early humans became better at providing for others in their group. The ability to hunt and collect surplus food and to support those unable to collect foods for themselves would have changed social life for early human species. The evolutionary trend for babies in the genus *Homo* to be born earlier meant that they needed more care, which in turn meant females needed additional provisioning. Not only did tool use influence the biological evolution of humans, it also affected their cultural evolution.

RELATED TOPICS
See also
ANCESTORS & TIMESCALES
page 138

THE LEAKEY FAMILY
page 148

HUMAN EVOLUTION:
THE FUTURE
page 152

3-SECOND BIOGRAPHIES
JANE GOODALL
1934 –
British anthropologist and primatologist and first to observe chimpanzee dipping stick use in 1960

CAREL VAN SCHAIK
1953–
Dutch primatologist and first to describe orangutan tool use

30-SECOND TEXT
Isabelle De Groote

Among our non-human relatives, chimpanzees are the most skilled in making and using tools – notably dipping sticks.

EVOLUTION OF THE BRAIN

the 30-second evolution

3-SECOND THRASH
Humans have enormous brains that make us smart, but we are not smart enough to work out exactly why!

3-MINUTE THOUGHT
Brains are complex organs that are made up of separate regions; while these regions are connected they are important for different functions. For example, the cerebellum is important for motor control, while the frontal lobe is involved in decision-making and memory. Comparing the relative sizes of such regions in humans and other mammals could shed light on important factors in the evolution of the human brain.

Humans have exceptionally large brains. Since we shared a common ancestor with chimpanzees some 6 million years ago, the human brain has undergone an unprecedented size increase. The average human brain weighs around 1.3kg (2.8 lb) whereas the average chimpanzee brain is under 500g (1.1 lb). The rich fossil record of the human lineage can be used to determine the trajectory of this increase over the last 4 million years or so. By studying fossil skulls scientists know that Australopithecines, one of the earliest hominids, had a brain about the size of a modern chimpanzee. By the time of *Homo erectus*, 1.5 million years ago, brain size had doubled. As human evolution progressed to the present this remarkable increase continued, with the largest brains (approximately 1.5kg [3.3 lb]) occurring in the Neanderthals that inhabited Europe tens of thousands of years ago. While the exceptional cognitive ability of the human is clearly facilitated by a large brain, the specific evolutionary driving force behind the increase in brain size remains the subject of vehement debate. A few strong contenders have emerged; these include the emergence of language, tool production, living in social groups and even tactical deception – which can be used to gain advantage over others.

RELATED TOPICS
See also
TOOL USE BY HUMANS
& OTHER APES
page 140

EVOLUTION OF HUMAN
LANGUAGE
page 144

HUMAN EVOLUTION:
THE FUTURE
page 152

3-SECOND BIOGRAPHY
HARRY JERISON
1928–
American pioneer of palaeoneurology who developed the encephalization quotient in the early 1970s

30-SECOND TEXT
Chris Venditti

Some scientists suggest climate change was a factor in increasing brain size – humans needed bigger brains to cope with their unstable, changing environment.

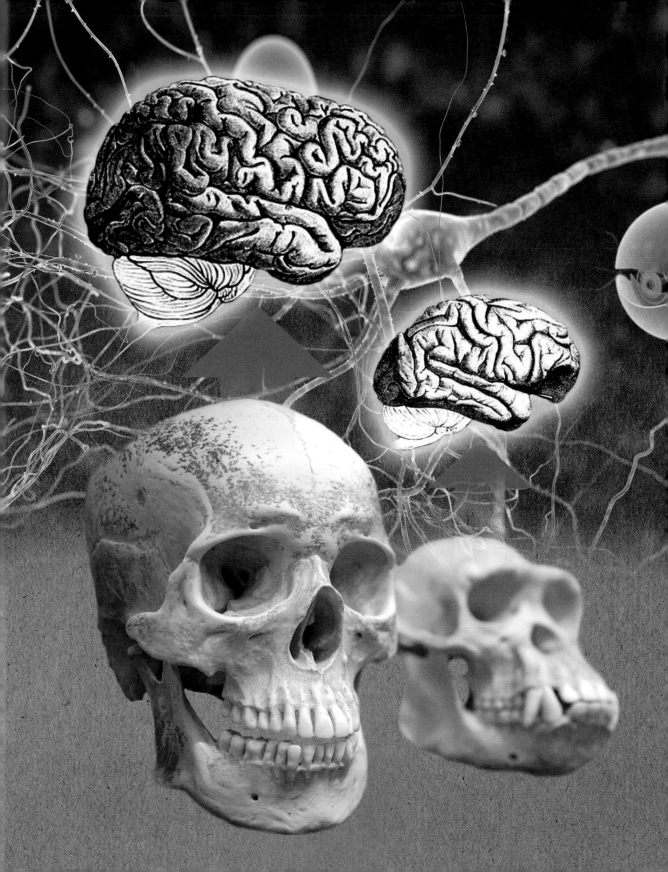

EVOLUTION OF HUMAN LANGUAGE

the 30-second evolution

3-SECOND THRASH
Human language evolved
to negotiate the increasing
complexity of human
society but without the
ability to understand what
is heard or spoken,
language would just be
empty words.

3-MINUTE THOUGHT
Where do the 6,000
human languages on
our planet come from?
Speech, hearing and brain
organization allowed
human ancestors to
improve the complexity
of their language: to
develop syntax, sense and
vocabulary as they spread
across the world and faced
different challenges and
environments. Despite the
present decline in global
language diversity, humans
will always find a way to
say what they need to say.

Research on the evolution of
language initially focused on symbolism and the
necessity to learn complex spoken language, but
recent work has shown that there are much
more basic requirements to human language
than the evolution of symbolic thought. In order
for language to exist, humans must be physically
able to speak and hear words. The earliest
evidence for anatomical changes associated
with spoken language is the enlarged spinal cord
in *Homo erectus*, which may have allowed
better breathing control for the production of
strings of words. Another change is seen in
Neanderthals where the hyoid bone, which
anchors the tongue, has the same semicircular
shape as that of humans. This suggests that
voice apparatus similar to that of humans was
present at least 450,000 years ago. Hearing
apparatus evolved, too: humans are the only
apes capable of hearing sounds around the
4 kilohertz frequency. This is the frequency
of many of the quiet consonants that are
important in conveying the meaning of words.
Studies of the fossil evidence of the inner ear
have shown that the ability to hear these quiet
consonants probably evolved with *Homo
heidelbergensis* as long as 1 million years ago.

RELATED TOPICS
See also
EVOLUTION OF THE BRAIN
page 142

EVOLUTIONARY PSYCHOLOGY
page 146

3-SECOND BIOGRAPHIES
PHILIP LIEBERMAN
1934–
American linguist studying the
biological evolution of
language

SUE SAVAGE-RUMBAUGH
1946–
American primatologist
studying ape language
capabilities

W. TECUMSEH FITCH
1963–
American evolutionary
biologist studying cognition
and communication in humans
and other animals

30-SECOND TEXT
Isabelle De Groote

*Language hardware ...
our ancestors had to
develop the physical
capacity to speak
and hear.*

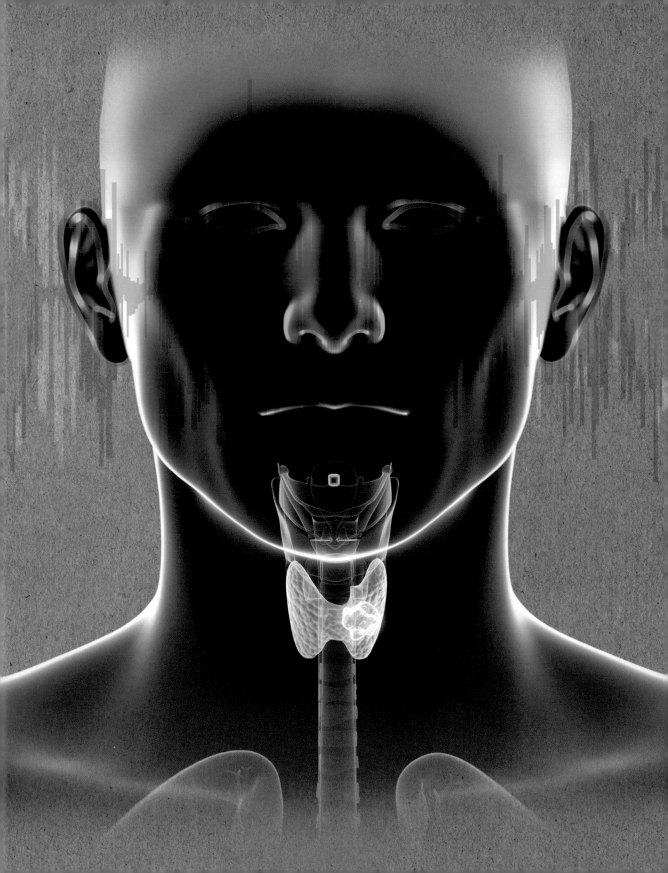

EVOLUTIONARY PSYCHOLOGY

the 30-second evolution

3-SECOND THRASH
Aspects of evolutionary
psychology date back to
Darwin, but it emerged
as a discipline in the
1970s, that aimed to
demonstrate how
evolution provides insights
into human behaviour.

3-MINUTE THOUGHT
Language is an example of
an important psychological
adaptation. Given that
humans universally learn
to talk without consistent
training, it seems likely
to be an evolved trait,
and there has been some
progress in identifying
genetic contributors to
language ability (though
unsurprisingly any attempt
to identify a specific
'language gene' has failed).
However, this view is
questioned by some
psychologists who believe
language to be an
accidental but beneficial
by-product of another
adaptation (a 'spandrel').

Although evolution was developed primarily as a theory with physical characteristics in mind, there is no reason why there should not be psychological traits or sexual preferences promoted by natural selection, as well as more obvious developments like eyes or wings. Many psychologists believe the mind is constructed from virtual modules not unlike physical organs and these cognitive modules – as well as characteristic behaviours – could evolve over time under the pressures of natural and sexual selection. Darwin wrote: 'In the future I see open fields for far more important researches. Psychology will be securely based on the foundation already well laid by Mr. Herbert Spencer, that of the necessary acquirement of each mental power and capacity by gradation.' The key developments bringing evolutionary psychology to prominence were Robert Trivers's concept of reciprocity, the basis of behaviour on a tit-for-tat response, and E.O. Wilson's work combining animal behaviour and social responses with evolutionary theory, leading to the recognition of evolutionary psychology as a separate discipline. These developments built on Bill Hamilton's suggestion that the gene was the driving force of evolution and behaviours that helped the survival of a person's genes in close relations would be reinforced selectively.

RELATED TOPICS
See also
FROM ADAPTATION TO
SPECIATION
page 48

ALTRUISM & SELFISHNESS
page 112

BILL HAMILTON
page 128

EVOLUTION OF HUMAN
LANGUAGE
page 144

3-SECOND BIOGRAPHIES
HERBERT SPENCER
1820–1903
British philosopher who used
the term 'survival of the fittest'

ROBERT L. TRIVERS
1943–
American biologist who
brought evolutionary
psychology to the fore

30-SECOND TEXT
Brian Clegg

What role has natural selection played in the development of how we think and how our minds work?

7 August 1903
Louis Leakey born in Kabete, British East Africa (now Kenya)

6 February 1913
Mary Nicol born in London

1928
Louis marries first wife Frida (mother of Colin)

1931
Louis's first expedition to Olduvai

13 December 1933
Colin Leakey born in Cambridge

1934
Louis leaves Frida for Mary Nicol

1936
Louis and Mary marry

4 November 1940
Jonathan Leakey born in Nairobi

28 July 1942
Meave Epps born in London

19 December 1944
Richard Leakey born in Nairobi

1948
Mary discovers Proconsul skull

1952
First extensive finds at Olduvai

1959
Skull of *Zinjanthropus* discovered

21 March 1972
Louise Leakey born in Nairobi

1 October 1972
Louis Leakey dies in Nairobi

1978
Mary excavates the Laetoli footprints

9 December 1996
Mary Leakey dies in London

1999
Meave's team discovers 3.5 million-year-old skull at Lake Turkana

Science occasionally produces a dynasty – for example, the father-and-son Nobel Prize-winning team of William and Lawrence Bragg – but there is something unique about the Leakey family, which includes palaeoanthropologists Louis Leakey and his wife Mary, their children plant scientist Colin, palaeoanthropologists Richard and Jonathan, Richard's palaeoanthropologist wife Meave and their daughter, Louise, a palaeontologist.

The family has never limited itself to science. Louis was involved in Kikuya tribal politics in Kenya from the late 1920s, Richard formed a political party called Safina in 1995 and became Kenya's Cabinet Secretary, and Jonathan briefly dabbled in palaeoanthropology before moving to business, running a snake venom company. But the Leakeys' main focus has always been the origins of humans in the study of ancient hominid remains. Together they have made a hugely significant contribution to our understanding of human evolution in Africa.

The large numbers of fossils the Leakeys excavated, particularly around the Olduvai Gorge in the Serengeti and Lake Turkana in Kenya, were fundamental to establishing the revolutionary idea that humans first evolved in Africa. They discovered early stone tools, chipped from stone that originated a number of miles away, suggesting their makers had significant mental capabilities. Mary's discovery of the remains of *Zinjanthropus bosei* (now *Paranthropus*), dating back 1.75 million years, resulted in a transformation of the accepted timescale of human evolution. Soon after, Jonathan found the first fragment of what would become known as *Homo habilis*, the earliest species to be (debatably) placed within the genus *Homo*.

Beyond their work with fossils, Louis helped set three of the world's leading primatologists on their paths – Jane Goodall, Dian Fossey and Biruté Galdikas – because he felt that there were similarities between the environment inhabited particularly by the great apes and that of a possible common ancestor of the great apes and humans, *Proconsul*. This genus of primates dating back more than 20 million years had first been identified by Arthur Hopwood when working with Louis; the first *Proconsul* skull was discovered by Mary in 1948.

Towards the end of Louis's life, he and Mary had professional disputes, particularly over his theory that humans arrived in the Americas around 100,000 years earlier than thought. Mary's work continued after Louis's death in 1972, and among her remarkable finds was the Laetoli footprints, excavated in 1978 at a site 45 kilometres (28 miles) from Olduvai. Around 3.6 million years old, the fossilized tracks from three individuals, preserved in volcanic ash, were then the oldest evidence for bipedal locomotion in a hominid. Meave and daughter Louise continue the Leakeys' six-decade long involvement with palaeontological research in Kenya.

Brian Clegg

HUMANS CAUSING EVOLUTION

the 30-second evolution

All organisms adapt to their changing environment, a process that leads to the evolution of species by natural selection. But around 10,000 years ago a new type of selection was introduced – human selection. At this time, humans exchanged their hunter-gatherer lifestyle for a more sedentary existence based on farming. A key element in this process was selection: of the best grasses with the biggest seeds (which became modern cereals like wheat and barley); and of the larger, more productive animals. Another aspect of this domestication process was breeding, as humans learnt to make crosses and select the best progeny for their own purposes. Early on these revolved around food supply, but more recently have come to include biofuel production and the generation of pharmaceutical compounds. Humans have made themselves agents of artificial selection and in the process have become an evolutionary force to rival natural selection. But by altering and shaping our environment, we have also accidentally driven the evolution of species. This is most obvious with the evolution of resistance of bacteria to antibiotics; of plants to herbicides; of insects to insecticides; and of rodents to rodenticides.

RELATED TOPIC
See also
HUMAN EVOLUTION:
THE FUTURE
page 152

3-SECOND BIOGRAPHIES
DANIEL ZOHARY
1926–
Israeli botanist who has studied genetic diversity in cultivated and wild cereals in the Fertile Crescent

GORDON HILLMAN
English archaeobotanist whose work includes studies of prehistoric cultivation and domestication of plants

30-SECOND TEXT
Isabelle De Groote

3-SECOND THRASH
Natural selection operating over millions of years has produced the diversity of organisms in the living world; humans, operating over a few thousand years have made themselves ever more powerful agents of evolutionary change.

3-MINUTE THOUGHT
Human-driven changes of traits in crops and domesticated animals show what we can achieve by artificial selection. In recent decades we have extended evolutionary novelty by introducing genes from unrelated species and designing new species in the laboratory. Encounters between laboratory species, artificially selected domesticates and naturally evolved organisms will create novel (and unpredictable) outcomes in the future.

The first farmers began human-driven evolution, which we have continued through many means – including GM crops and cloning of animals.

HUMAN EVOLUTION: THE FUTURE

the 30-second evolution

3-SECOND THRASH
Evolution by natural selection made humanity; evolution by culture will complete it.

3-MINUTE THOUGHT
By allowing those to live who would previously have died because of genetic predispositions, our species is evolving but not by Darwinian natural selection, as those genes are not removed from the gene pool. By 2050 will there be 9 billion people of greater genetic diversity than ever before?

Global population increases by over 100 people every minute. This massive change influences everything on the planet, including our own evolutionary future. Take a recent example of human evolution: the ability of adults to digest milk resulted from a mutation in a gene that normally switches off the production of lactase as we are weaned. When that first arose around 10,000 years ago, individuals in small European populations who possessed the mutation could take advantage of the abundant milk produced by newly domesticated cows, with the gene sweeping rapidly across connected populations. This would probably not happen now, as evolution relies upon isolation and small populations, and the increasingly connected nature of the world means that we are undoing the local patterns of adaptation that have arisen over time. And we are on the cusp of an even greater change: should society allow it, genetic engineering technology will allow parents to manipulate their children's genetic fate. Humankind is an extraordinary product of natural selection, but we are moving beyond Darwin's understanding of survival of the fittest.

RELATED TOPIC
See also
HUMANS CAUSING EVOLUTION
page 150

30-SECOND TEXT
Mark Fellowes & Nick Battey

What is the future of our evolution? Are we moving beyond natural selection – and 'survival of the fittest' – to synthetic selection in a culturally determined gene pool?

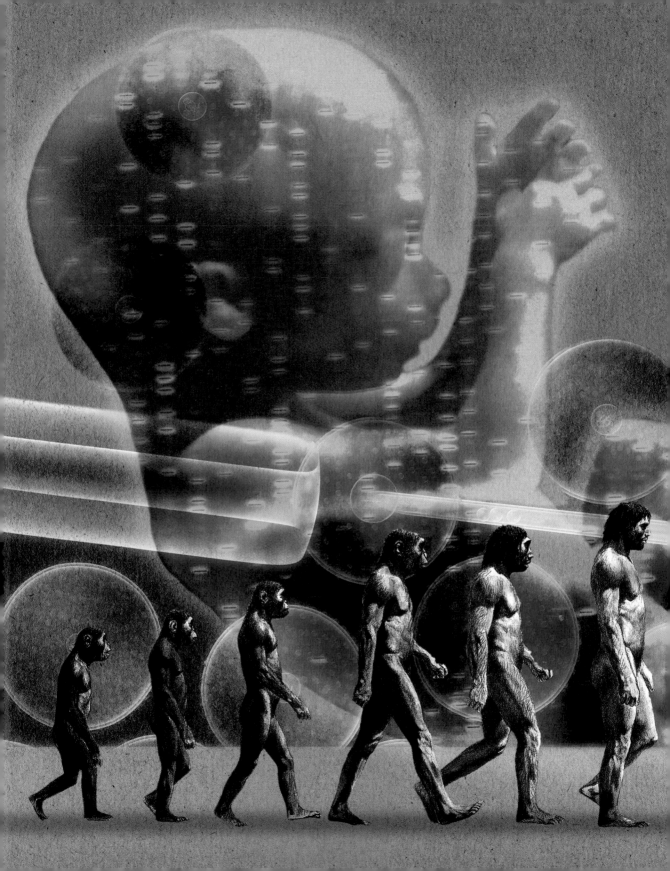

RESOURCES

BOOKS

99% Ape: How Evolution Adds Up
Jonathan Silvertown
(Natural History Museum, 2008)

*The 10,000 Year Explosion: How
Civilization Accelerated Human Evolution*
Gregory Cochran & Henry Harpending
(Basic Books, 2011)

*Chimpanzee Material Culture: Implications
for Human Evolution*
William Clement McGrew
(Cambridge University Press, 1992)

Darwin's Dangerous Idea
Daniel C. Dennett
(Penguin, 1996)

*The Darwinian Revolution: Science Red
in Tooth and Claw*
Michael Ruse
(University of Chicago Press, 1999)

Evolution: The History of an Idea
Peter J. Bowler
(University of California Press, 2009)

*Female Control: Sexual Selection by
Cryptic Female Choice*
William G. Eberhard
(Princeton University Press, 1996)

*The Human Story: Where we Come From and
How we Evolved*
Charles Lockwood
(Natural History Museum, 2013)

*Lone Survivors: How we Came to be the Only
Humans on Earth*
Chris Stringer
(Griffin, 2013)

*Nature's Nether Regions: What the Sex Lives
of Bugs, Birds and Beasts Tell Us About
Evolution, Biodiversity and Ourselves*
Menno Schilthuizen
(Penguin, 2015)

The Selfish Gene (30th anniversary edn.)
Richard Dawkins
(Oxford University Press, 2006)

*Sexual Selection and the Origins of Human
Mating Systems*
Alan F. Dixson
(Oxford University Press, 2009)

*Sexual Selections: What We Can and Can't
Learn about Sex from Animals*
Marlene Zuk
(University of California Press, 2003)

The Story of the Human Body: Evolution, Health and Disease
Daniel Lieberman
(Vintage Books, 2014)

Tool Use in Animals: Cognition and Ecology
Crickette Sanz, Josep Call &
Christophe Boesch
(Cambridge University Press, 2013)

Why Evolution is True
Jerry Coyne
(Oxford University Press, 2010)

WEBSITES

www.newscientist.com/topic/evolution
Articles and features to cover all areas of
evolution, from introductions to modern
findings and current theories.

darwin200.christs.cam.ac.uk
The University of Cambridge web page that
celebrates the life, work and impact of
Charles Darwin.

www.nhm.ac.uk/nature-online/evolution/
how-did-evol-theory-develop/
The Natural History Musuem, London, web
page, tracking the development of Evolution
as a theory.

www.humanorigins.si.edu
The Smithsonian's National Museum of
Natural History web page exploring the latest
findings and implications on the scientific
exploration of human origins.

https://genographic.nationalgeographic.com
The Genographic Project is a multiyear
research initiative led by National Geographic
Explorer-in-Residence Dr. Spencer Wells. It's
main focus is to analyze historical patterns in
DNA from participants around the world to
better understand human genetic roots.

NOTES ON CONTRIBUTORS

EDITORS

Nick Battey is Professor of Plant Development at the University of Reading. He has published extensively on pure and applied plant biology and has a strong interest in the history of biology. He is co-author of the book *Biological Diversity: Exploiters and Exploited*. Nick completed a BSc in Plant Science at the University of Wales and a PhD in Plant Developmental Biology at Edinburgh University. He is currently Head of the Ecology & Evolutionary Biology Section in the School of Biological Sciences.

Mark Fellowes is Professor of Ecology at the University of Reading. His broad research interests span questions which range from how insects evolve resistance to their natural enemies through to the consequences of urbanization on the abundance and diversity of wildlife. He was lead editor of the book *Insect Evolutionary Ecology*. Mark completed his BSc in Zoology and his PhD in Evolutionary Biology at Imperial College London before moving to Reading, where he is currently Head of the School of Biological Sciences.

CONTRIBUTORS

Brian Clegg read Natural Sciences, focusing on experimental physics, at the University of Cambridge. After developing hi-tech solutions for British Airways and working with creativity guru Edward de Bono, he formed a creative consultancy advising clients ranging from the BBC to the Met Office. He has written for *Nature*, the *Times*, and the *Wall Street Journal* and has lectured at Oxford and Cambridge universities and the Royal Institution. He is editor of the book review site www.popularscience.co.uk, and his own published titles include *A Brief History of Infinity* and *How to Build a Time Machine*.

Isabelle De Groote is a senior lecturer in Biological Anthropology at Liverpool John Moores University and Scientific Associate at the Natural History Museum, London. She has published research on human evolution ranging from the limb bones of the Neanderthals, to the earliest human footprints in Britain. Isabelle has also written popular articles for *BBC Focus* Magazine and has been a scientific consultant on a number of television documentaries on human evolution.

Julie Hawkins is Associate Professor of Plant Systematics and Evolution at the University of Reading. The academic road to Reading took her via King's College, Univeristy of London, and the Universities of Birmingham, Oxford and Cape Town. She has published more than forty peer-reviewed research papers on topics ranging from cactus conservation to convergent evolution of floral form. She has a particular enthusiasm for definitions of homology. A passionate communicator, she also enjoys the opportunities to teach that her role affords her.

Louise Johnson is a lecturer in Population Genetics at the University of Reading. Her research is on the evolution of genetic systems: sex, genetic codes, gene regulatory networks and cancer. She first became involved in science communication through blogging, contributing to the anthology *Open Laboratory 2006*, and later via the Wellcome Trust's schools outreach event *I'm A Scientist*. She was a guest speaker at ZSL-L'Oreal's 2014 SoapBox Science showcase in London.

Ben Neuman has researched deadly viruses on three continents, and in the process has probably grown more SARS virus than any living person. He currently enjoys tinkering with the fascinatingly cryptic machinery of life in his laboratory at the University of Reading, where he also indulges his passion for fossils of the earliest animal life. Ben has authored over 30 scientific papers on topics ranging from how viruses evolve to what makes them tick.

Chris Venditti is an evolutionary biologist at the University of Reading. His research seeks to identify large scale patterns and processes of evolution that have occurred over geological time scales, using a variety of data sources from molecular sequences to the fossil record. He has published scientific papers on speciation, molecular evolution, phenotypic evolution and adaptation.

INDEX

ACKNOWLEDGEMENTS

A special thank you to James Hunter for his knowledge and expertise.

PICTURE CREDITS
The publisher would like to thank the following individuals and organizations for their kind permission to reproduce the images in this book. Every effort has been made to acknowledge the pictures; however, we apologize if there are any unintentional omissions.

All images from Shutterstock, Inc./www.shutterstock.com and Clipart Images/www.clipart.com unless stated.

David Scharf/Science Photo Library: 47.

Martin Shields/Science Photo Library: 61.

Bettmann/Corbis: 62.

Dr Marli Miller/Visuals Unlimited, Inc./Science Photo Library: 91

The Natural History Museum/Alamy: 101.

James King Holmes/Science Photo Library: 128.

Stock Connection Blue/Alamy: 127.

Bettmann/Corbis: 148.

Getty Images/Handout: 151.

David Gifford/Science Photo Library: 153.